GENERAL CHEMISTRY AS A SECOND LANGUAGE

DR. DAVID R. KLEIN

Johns Hopkins University

JOHN WILEY & SONS, INC.

ACQUISITIONS EDITOR	Debbie Brennan
MARKETING MANAGER	Amanda Wygal
PRODUCTION MANAGER	Pamela Kennedy
PRODUCTION EDITOR	Sarah Wolfman-Robichaud
SENIOR DESIGNER	Kevin Murphy
SENIOR ILLUSTRATION EDITOR	Anna Melhorn
PROJECT MANAGEMENT SERVICES	Pam Lininger/Matrix Publishing Services
COPYEDITOR	Kathy Drasky

This book was set in 10/12 Times Roman by Matrix Publishing and printed and bound by Courier Westford. The cover was printed by Phoenix Color.

This book is printed on acid-free paper.∞

ISBN 0-471-71662-6

Printed in the United States of America

10 9 8 7 6 5 4 3 2 1

INTRODUCTION

HOW TO USE THIS BOOK

Without mastering certain fundamental skills, it is very difficult to do well in a chemistry course. I refer to these basic skills as the "language" of chemistry. Students *MUST* learn how to convert units, balance reactions, calculate enthalpy values, and so forth. This book provides explanations and related exercises geared toward fine-tuning these essential skills.

This book **cannot** be used to replace your textbook, lectures, or other studying. This book is **not** the *Cliff Notes* of chemistry. Rather, it is meant as a resource for you to acquire the skills that will enable you *to study more efficiently*. This book will not teach you how to solve *every* kind of problem. Rather, it will provide you with the *core language* that will help you *build your confidence* in the course. This will then empower you to study from your textbook more efficiently. This book is meant to be a companion to any general chemistry textbook.

It was very difficult to select the concepts considered to be the basic language of chemistry. Some topics, in the interest of length restrictions, are not included. Topics were prioritized, and some had to be cut. In the end, the topics covered are selected topics from the *first semester* only. For example, topics such as acid-base reactions, equilibrium, and Gibbs free energy are often taught in either the first or second semester (depending on the instructor). Therefore, these topics are not included in this book.

How to Use This Book

In order to best use this book, you need to know the **two major aspects** to study in this course.

How to Study

There are **two** separate aspects to this course:

1. understanding principles
2. solving problems

Although these two aspects are completely different, instructors will typically assess your understanding of the principles by testing your ability to solve problems.

In this book, we will look at some common types of problems, and we will fine-tune your basic skills.

The only way to truly master problem solving is to practice problems every day, consistently. You will never learn how to solve problems by just reading. You must try, and fail, and try again. You must learn from your mistakes. You must get frustrated when you can't solve a problem. That's the learning process. The worst thing you can do is to read through the solutions manual and believe that it adequately prepares you to solve problems on an exam. It doesn't work that way. If you want an A, you will need to sweat a little (no pain, no gain).

The simple formula: review the principles and concepts; then *focus all of your remaining time on solving problems*.

TABLE OF CONTENTS

CHAPTER 1

NUMBERS AND UNITS

If you flip through your textbook and look at the kinds of problems you will be solving in each chapter, you will find that the answer to almost every problem is a number. You might see an occasional question asking you to explain some principle or concept, but in general, most problems require you to *calculate* an answer. Whether it is the pH of a solution or the enthalpy of formation for a reaction, you will be spending a lot of time learning how to use equations to calculate answers. But even if you know what equations to use and you know how to calculate the answer, it is still possible to get the answer wrong.

There are TWO equally important parts of any answer: (1) the number, and (2) the units. Here are some simple examples:

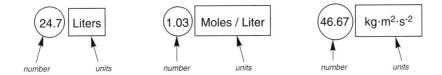

If you incorrectly report either one of these two parts, then you will mess up your answer. You need to learn how to report each part properly. This might seem like simple stuff, but little mistakes can ruin your answer (and you can lose precious points on an exam). So let's try to avoid that, and let's get some practice. Most of the problems in this chapter are very simple and won't take long to do, so you should be able to breeze through this chapter pretty quickly. In fact, much of the material in this chapter is just a review of what your intuition already knows. The main point of the chapter is to reinforce the *skills* you need to report an answer (so that you won't have to rely on your intuition during an exam).

In our discussion of **units** toward the end of this chapter, we will see a very important method that helps us ensure that our answers have the correct units. This method, **the factor-label method**, will be used throughout the entire course, so we need to *master* this method early on.

We will start our discussion with the first part of every answer: the number.

1.1 SIGNIFICANT FIGURES

To understand the importance of significant figures, consider the following example. The current estimate for world population is *about* 6.5 billion. Notice the im-

1

portant term "about". No one really knows the exact number, so we must use an estimate. Now imagine dividing up the world into three regions with equal populations. The number of people in each region is just 6.5 billion divided by 3. If you key that into your calculator, you will get 2,166,167 people. Even though the calculator gives this exact answer, we know that each region will NOT have *exactly* that number of people, because the starting number (6.5 billion) was just an estimate. So, when we divide by 3, our answer must also be an estimate. The starting number had only two significant digits (the 6 and the 5), so our answer cannot have more than two significant digits either. Our answer must be 2.2 billion people in each region.

If we want to have an answer with more significant figures, then we will have to start with an estimate that has more significant figures. For example, if we learn that the newest estimate for world population is 6.532 billion, then we have a starting number with four significant figures. So, our answer (6.532 billion divided by 3) will also have four significant figures. If you key that into your calculator and round your answer to the fourth digit, your answer will be 2.177 billion.

Now comes the important part—if the answer you write on an exam has too many OR too few significant figures, then the answer is WRONG. If you write down 2.1773333 billion instead of 2.177 billion, then the answer is wrong, *even if the calculator says 2.1773333*. First we will learn how to count significant figures, and then we will learn how to determine how many significant figures belong in an answer.

Here is the way to count significant figures: Start at the left side of the number and move toward the right until you get to the first non-zero digit. Start counting at that digit. Examples:

Once you figured out where to start counting, then you have to find out where to end counting. If the number has a decimal in it, then you count until the very end:

If the number does *not* have a decimal, then we generally count until the last non-zero number:

This last set of examples points out a difficulty. Suppose you estimate the number of people at a baseball game to be *approximately* 8500. Your number is just an estimation (there could be 8527 or 8496 people). The way you report your number indicates that you have estimated the number to the nearest hundred—that's why you show two significant figures in your number (the 8 and the 5). If you had rounded to the nearest thousand, you might have estimated 9000 people (which would be just one significant number). However, what if you count each and every person at the baseball game and the number comes out to be exactly 8500 people? How do you report that your number is exact and not just an estimation? If you just say 8500, everyone will assume that you are talking about *two* significant figures. We will show how to solve this issue in the next section. For now, we will consider the number 8500 as having two significant figures.

Let's get practice counting significant figures, so that we can get comfortable using them properly in our answers.

EXERCISE 1.1. Count the significant figures that you see in the number 0.007520.

Answer: Look at the far left of the number, and move right until you see the first non-zero number. That's where you start counting. In this case, you start counting with the 7:

Next you have to figure out how far to count. In this case, there is a decimal point, so you count until the very end:

As you can see, there are four significant figures.

Let's get a little bit of practice before we move on. For each of the following problems, how many significant figures do you see? Remember that you always start with the first non-zero digit, and you decide where to stop counting based on whether there is a decimal. If there is a decimal, count all the way to the end of the number. If there is no decimal, count until the last non-zero digit.

1.2. 0.0713200   **1.3.** 7843000   **1.4.** 1.4800

1.5. 100   **1.6.** 100.0   **1.7.** 894.003

1.8. 89400   **1.9.** 0.03000   **1.10.** 74.000

Now that we have seen how to count significant figures, we will use that skill to multiply and divide two or more numbers when one of the numbers has less significant figures than the other(s). For example, if you had to multiply 0.034 and 127, you should notice that 0.034 has two significant figures and 127 has three significant figures. When it comes to multiplying and dividing, the rule goes like this: look at the number with the fewest significant figures, and your answer should have the same amount of significant figures. In our case, the number with the fewest significant figures is 0.034, which has only two significant figures. So, in this example, your answer should have only two significant figures. If you use your calculator to multiply 0.034 and 127, you will get 4.318. But you must fight the temptation to write that down as your answer. You must round this number to only two significant figures. So the answer is 4.3.

The same rule applies when you divide two numbers. You just choose the number with the fewest significant figures, and that is how many significant figures should be in your answer.

In each of the following problems, determine how many significant figures should be present in the answer:

1.11. 472×101 **1.12.** 4600×0.005 **1.13.** 36.0×4752

1.14. $\dfrac{45.08}{36.2}$ **1.15.** $\dfrac{1.003}{8500}$ **1.16.** $\dfrac{0.003}{472} \times 12$

1.17. $\dfrac{3.003}{475.0} \times \dfrac{0.30}{524}$ **1.18.** 0.3005×4.1

So far, we have seen how to multiply and divide numbers that have a different amount of significant figures. Now we need to see how to add and subtract, because the rules are a bit different. For adding and subtracting, we count the decimal places. For example, 472.44 has two decimal places (there are two digits appearing after the decimal); 890 has no decimal places. Look at the number with the fewest decimal places, and make sure that your answer has the same amount of decimal places. So, if you add 45.62 and 3.7, your calculator will say 49.32. But don't write that down as your answer. One of the starting numbers (3.7) only had one decimal place, so your answer must also have only one decimal place. Thus, round your answer off to 49.3. Similarly, if you are adding 45.62 and 24, then your answer should be 69 (with no decimal at all), because 24 has no decimal places.

The same rule applies when subtracting numbers. You just choose the number with the fewest decimal places, and that is how many decimal places should be in your answer.

In each of the following problems, determine how many decimal places should be present in the answer:

1.19. $23.56 + 24.983$ **1.20.** $4.78 - 2.892$ **1.21.** $46.83 - 0.03$

1.22. $34.892 + 5.0$ **1.23.** $134.033 - 0.02$ **1.24.** $48.2 - 46$

So far, we have seen the rules to determine how many digits to put in our answer. But to actually write down the answer, you will need to round the answer displayed on your calculator screen, so that the answer you write has the correct number of digits. For the most part, rounding is intuitive, but there is one part that is definitely not intuitive. We'll explore this now.

When rounding off a number, we look at the digit that is directly beyond the last significant digit. For example, if the answer on your calculator is 34.27, and you have determined that you must round off to three significant figures, then you look at the fourth digit and ask if it is more than 5. In this case (34.27), the 7 is the first digit that is not significant. Since 7 is more than 5, we round up, and round off our answer to 34.3. Similarly, if the first nonsignificant digit is less than 5, then we round down. For example, if your calculator says 45.782, and you have determined that you can only have 4 significant figures, then look at the fifth digit (which is a 2) and round down to 45.78.

Now you know what to do if your first nonsignificant digit is more than 5 or less than 5. But what if it is exactly 5? Then what? It's easy if something comes after that 5. For example, if we have the number 43.501 and we have to round off to 2 significant figures, then we would round up and our answer is 44. But what about a case where nothing comes after the 5? Here is the part where it is not so intuitive. Your intuition probably tells you to always round up. For example, you might say that 43.5 should be rounded up to 44, and 44.5 should be rounded up to 45. But chemists have come up with a different way of doing things. It goes like this: When you have a 5 right after the last significant figure, then look at the last significant figure, and ask whether it is odd or even. In the case of 43.5, the digit before the 5 is odd. In the case of 44.5, the digit before the 5 is even. Believe it or not, the convention is to treat these differently. We round UP when the last significant figure is odd, but we round DOWN when the last significant figure is even. That's right. As crazy as it might sound, 42.5 will get rounded DOWN to 42, because the last significant figure (2) is an even digit. We treat 0 as an even number, so 40.5 would get rounded DOWN to 40.

Let's get some practice. Using the rules you just learned, round off the following numbers to two significant digits:

1.25. 34.78 _____ **1.26.** 24.33 _____

1.27. 17.51 _____ **1.28.** 17.50 _____

1.29. 18.50 _____ **1.30.** 20.5 _____

1.31. 45.50001 _____ **1.32.** 45.5000 _____

We have now seen a lot of rules, and we are ready to put them all together to do some calculations. Let's just quickly review everything we have seen:

1. When counting significant figures, start from the left at the first non-zero number. If there is a decimal in the number, then count until the end. If the there is no decimal, then count until the last non-zero number.

2. When multiplying or dividing numbers, choose the number with the fewest significant figures, and that is how many significant figures should be in your answer.

3. When adding or subtracting numbers, choose the number with the fewest decimal places, and that is how many decimal places should be in your answer.

4. When rounding off an answer, look at the first nonsignificant number. If it is less than 5, round down. If it is more than 5, round up. If it *is* 5, then look to see if there are any non-zero digits after the 5. If so, round up (e.g., 42.500001 will round up to 43). If there are no non-zero digits after the 5, then you must look at the digit before the 5—round up if odd; round down if even (e.g., 42.50000 will round down).

EXERCISE 1.33. Make the following calculation and report your answer with the correct amount of significant figures.

$$8.410 \times 5.00 = \underline{\hspace{2cm}}$$

Answer: Begin by counting the significant figures in each of the starting numbers. In the first number (8.410), begin with the 8. This number has a decimal, so count all the way until the end, including the zero. You'll note that it has *four significant figures*. In the second number (5.00), begin with the 5. This number also has a decimal, so by counting until the end, including both zeros, you see that there are *three significant figures*. Therefore, your answer should have three significant figures.

Now multiply the numbers on your calculator: 8.410×5.00. Your calculator should say 42.05. We just determined that we have to round off to three significant figures, so we look at the fourth digit to see whether to round up or down. The fourth digit is a 5, and this is situation where it is not so intuitive so we have to use the rules. Look back to the last significant digit, which is a zero. Remember that a zero counts like an even number, and when the last significant digit is an even number, then the 5 gets rounded DOWN. So, the answer is 42.0.

Now let's get some practice. For each of the following mathematical operations, calculate the answer and record your answer with the correct number of significant digits.

1.34. $34.03 \times 0.0072 = \underline{\hspace{2cm}}$ **1.35.** $\dfrac{3.003}{475.0} = \underline{\hspace{2cm}}$

1.36. $67.75 - 7.2 = \underline{\hspace{2cm}}$ **1.37.** $356.50 + 6.0 = \underline{\hspace{2cm}}$

1.38. $356.50 + 12 = \underline{\hspace{2cm}}$ **1.39.** $356.50 + 13 = \underline{\hspace{2cm}}$

1.40. $\dfrac{3.003}{475.0} \times 0.0322 =$ _____ **1.41.** $1.003 \times 5.0 =$ _____

1.42. $46.7 - 0.4 =$ _____ **1.43.** $\dfrac{12.4}{1.7} =$ _____

Before we move on to the next section, there are two important points to make:

1. When you are making a series of calculations, you should wait to round off your answer until the very end. Do all of your calculations without any rounding off, and then just round once at the end. If you are working on a long problem and you do round your answers after each calculation in a series of calculations, then you might find that the last digit of your answer might be different from the answer in the back of your textbook.

2. Whenever you count discrete objects, significant figures don't apply. For example, if you count that there are 6 chairs around the table, then the number 6 is an exact number. It is not an estimate. We don't treat this number as if it has one significant figure. Rather, we treat it as if it has an infinite number of significant figures. For example, if you know that each chair weighs 2.34 pounds, and you want to know how much 6 chairs will weigh, then you need to multiply 6 times 2.34. In order to determine the amount of significant figures to put in your answer, look at the number with the fewest significant figures. When looking at the two numbers, 6 and 2.34, we would normally say that 6 has the fewest significant numbers. But in this case, the number 6 is counting discrete objects, so it is treated as if it has an infinite number of significant digits (6.00000 . . .). So, in this case, the number with the fewest significant figures will be 2.34. Therefore, your answer should have three significant figures in it.

1.2 ORDERS OF MAGNITUDE

Most of the numbers that you will use in this course are either very large or very small. It gets a little bit crazy to talk about numbers like 4,782,000,000,000,000,000,000 molecules. Similarly, it is crazy to talk about numbers like 0.0000000000843 grams. There are so many zeros in each of these numbers that it is impossible to read them and get a sense of their magnitude. All we can say is that the first number is very big and the second number is very small. To get a real sense of *how* big and *how* small, we express numbers in *scientific notation*—a system based on powers of 10. You have probably seen this system already in high school, but let's just review quickly. Scientific notation is expressed like this: 10^1, 10^2, 10^3, 10^4, etc. The superscript above the 10 tells us the magnitude of the number. So,

$$10^2 = 100$$
$$10^3 = 1000$$
$$10^4 = 10,000$$
$$\text{and } 10^{23} = 100,000,000,000,000,000,000,000$$

If we want to express a number like 4,782,000,000,000,000,000,000, then we can just say 4.782×10^{21}.

In a moment we will review how to convert numbers into scientific notation, we will revisit significant figures as applied to scientific notation, and we will see the subtleties of performing arithmetic operations on these numbers (multiplying and dividing). But first, we should take a moment to reflect on the magnitude that is conveyed when we use powers of 10.

As an example, consider the age of the universe—13.7 billion years. Try to guess how many seconds are in 13.7 billion years. 10 to the power of what? 10^{100}? 10^{1000}? Would you believe that it is just 10^{17}? Just multiply the following numbers: (13.7 billion) $\times$ (365) $\times$ (24) $\times$ (60) $\times$ (60) and your calculator will give you the answer: 432,043,200,000,000,000. When you convert this answer into scientific notation and express the appropriate significant figures (we will see how to do this very soon), you get 4.32×10^{17} seconds. Now we can begin to appreciate that 10^{17} is a REALLY BIG number. We cannot even begin to relate to a number that large. But consider this: 10 times that number is 10^{18}, and 10 times that is 10^{19}, and 10 times that, and 10 times that, and 10 times that, and 10 times that—is roughly the same number as the number of molecules in a teaspoon of water (10^{23}). *Now that is truly mind boggling!* Clearly, it would be a gross understatement to say that there are billions of water molecules in that teaspoon of water. That would be like saying that Bill Gates is worth a few pennies. The magnitude would be totally off.

The *order of magnitude* (the exponent on the 10) tells us how big or small a number is. Compare 10^9 and 10^6. 10^9 is three orders of magnitude larger than 10^6. That means that 10^9 is 1000 times larger than 10^6. It is important to realize the scope of these numbers, because many numbers in chemistry rely on your ability to appreciate orders of magnitude. Exponents are often used throughout the course, and so are logarithms (logarithms are just a fancy way of repackaging exponents). The last chapter (Chapter 9) of this book provides a more detailed explanation of exponents and logarithms.

The problems that you will solve in your chemistry course will often have answers that must be expressed using exponents. Students often get confused when we start manipulating exponents. So, we need to get some practice converting regular numbers into numbers with exponents, and vice versa. Here's how we do it.

Consider the number 362. We know that we can get this number if we multiply 36.2×10. We also get the same number if we multiply 3.62×100. Notice that we have moved the decimal place so that there is now only one digit to the left of the decimal. This is exactly how far we will always move the decimal. Now we express the number in terms of powers of 10, like this: (3.62×100) is the same as 3.62×10^2. When we express a number in this format, we call it **scientific notation**.

So we see that the procedure for converting a number into scientific notation is fairly simple. We count how far we need to move the decimal place so that there will be one non-zero digit to the left of the decimal. Then we multiply by a power of 10 that tells us how far we moved the decimal. Let's see an example:

EXERCISE 1.44. Express the following number in scientific notation: 400374.2.

Answer: Begin by counting how far to the left you need to move the decimal in order for there to be exactly one digit to the left of the decimal:

$$4\,0\,0\,3\,7\,4.0$$

You'll see that you need to move the decimal place 5 times to the left. If you did not multiply by a power of 10, then you would be making the number much smaller by moving the decimal over 5 places. So, to compensate, multiply the new number by 10^5. Your answer is: $\mathbf{4.003740 \times 10^5}$. Make sure that you don't drop the last zero. It is a significant figure, so it must stay in your answer.

For each problem below, express the number in scientific notation:

1.45. $1435.27 = $ _____ **1.46.** $243.1 = $ _____

1.47. $3801 = $ _____ **1.48.** $274.30 = $ _____

Sometimes, you have a number where there is no decimal place. In cases like this, just place a decimal at the end of the number, and then count how far you have to move the decimal place. Let's see an example.

EXERCISE 1.49. Express the following number in scientific notation: 4642.

Answer: Place a decimal at the end of the number, and then count how far to the left you need to move the decimal in order for there to be exactly one digit to the left of the decimal:

$$4\,6\,4\,2$$

You'll see that you need to move the decimal to the left 3 times. So, you need to multiply the new number by 10^3. And your answer is: $\mathbf{4.642 \times 10^3}$.

For each problem below, express the number in scientific notation:

1.50. $243 = $ _____ **1.51.** $372{,}567 = $ _____

1.52. $28{,}942 = $ _____

Now we can answer a question asked in the previous section when we were discussing significant figures. We considered a number like 8500. And we wanted to know how many significant figures it has. If we use the rules that were laid out, we would find that there are two significant figures (the 8 and the 5). But what if you are counting people at a baseball game and you find that there are exactly 8500? How do you indicate that this is not an estimate, and that the zeros are significant figures? If we use scientific notation, this issue goes away. If we want to say that 8500 is just an estimate, rounded to the nearest hundred, then we would say 8.5×10^3. Notice that there are only two significant figures here. But if we want to show that 8500 is an exact number and the zeros are significant, then we would say 8.500×10^3. Notice that this number shows that all four digits are significant. In scientific notation, counting significant figures is easy. You just count every digit. If you go back to the rules outlined in the previous section, you will see why. In scientific notation, the first digit will always be the first significant digit (by definition), and scientific notation always has a decimal place (so you always count all digits until you get to the end of the number).

For each problem below, assume that the zeros at the end are not significant, and express the number in scientific notation:

1.53.   4890 = _____   **1.54.**   240 = _____

1.55.   3700 = _____

So far, we have seen how to express a large number in scientific notation. Now, we turn our attention to expressing small numbers. Consider the number 0.000043. In order to move the decimal place so that there is one non-zero digit to the left of the decimal, you will have to move the decimal five times *to the right*.

$$0.0\,0\,0\,0\,4\,3$$

We express this by using a negative exponent: 4.3×10^{-5}. Let's see why. First we need to understand what a negative exponent means. A negative exponent is used for small numbers. We already saw that $1000 = 10^3$. So 1 / 1000 would just be 1 / 10^3. That is where negative exponents come in to the picture:

$$\frac{1}{10^3} = 10^{-3}$$

So, when we use a negative power of 10, we are just saying that number is a small number. In the example above, 4.3×10^{-5} is a small number.

If you want, you can memorize that moving the decimal place to the left gives a positive exponent, and moving the decimal place to the right gives a negative exponent. But this method will confuse you later on when you practice converting from scientific notation back into regular numbers. In that case, everything gets reversed, and you will get confused if you are just memorizing. So, it is much better to think about the size of the number. You can do this with the following example.

EXERCISE 1.56. Express the following number in scientific notation: 0.008620.

Answer: First count how far you need to move the decimal in order for there to be exactly one non-zero digit to the left of the decimal:

$$0.008620$$

You'll see that you need to move the decimal three times. So, you need to decide whether the exponent will be 3 or -3. Instead of memorizing which way you moved the decimal (left or right), compare your original number (0.008620) to your new number after you move the decimal (8.620). Now, ask yourself which number is larger. Clearly, the new number (8.620) is larger. But the new number can't be different than the old number. When we express a number in scientific notation, we are not changing the number; we are just writing it differently. So we must multiply a power of 10 that makes our new number (8.620) much smaller so that it will be the same as 0.008620. And so, the exponent needs to be negative, which means your answer is: $\mathbf{8.620 \times 10^{-3}}$. Don't forget to keep the zero at the end of the number so that you have the correct number of significant figures

For each problem below, express the number in scientific notation. In each case, you will have to decide whether the exponent should be positive or negative:

1.57. 0.000567 = _____ **1.58.** 2400.7 = _____

1.59. 370.24 = _____ **1.60.** 0.00000564 = _____

1.61. 0.00432 = _____ **1.62.** 0.03840 = _____

Now that you have seen how to convert regular numbers into scientific notation, the reverse process is just the opposite of this. Let's see an example.

EXERCISE 1.63. Consider the following number expressed in scientific notation: 4.634×10^{-3}. Express this number in regular notation.

Answer: Once again, do not memorize which way to move the decimal based on whether the exponent is positive or negative. If you try to memorize, you will get confused because you have to take into account whether you are going from scientific notation to regular notation or vice versa. Let's just figure it out based on the size of the number. 10^{-3} means that the number we are talking about is a small number. Therefore, you need to move the decimal three spaces to the left, which will give you **0.004634**.

Let's double check to make sure this was done correctly. If you had moved the decimal the other way, you would have gotten 4634. That can't be the answer because you can see from the negative exponent (4.634×10^{-3}) that the number is a small number.

For each problem below, express the number in regular notation. In each case, you will have to decide which way to move the decimal. Try to figure it out based on analyzing the relative size of the number:

1.64. $2.48 \times 10^3 =$ _____   **1.65.** $4.64 \times 10^{-3} =$ _____

1.66. $7.924 \times 10^{-2} =$ _____   **1.67.** $7.924 \times 10^2 =$ _____

1.68. $8.456 \times 10^{-4} =$ _____   **1.69.** $3.84 \times 10^5 =$ _____

So far, we have seen how to write numbers in scientific notation. Now, we need to see how to multiply numbers that are expressed in scientific notation. Suppose we need to multiply the following (3.45×10^2)(8.9×10^4). The way we do this is by adding the exponents (for a more detailed explanation of this, see Chapter 9), which will give us $3.45 \times 8.9 \times 10^{(2+4)}$. So, we have a power of 10 that is now 10^6. We just added the exponents (2 and 4). But there are a couple of things to be careful about:

- Don't forget about significant figures. You can't ignore everything you learned in the previous section. The first number (3.45) has three significant figures, and the second number (8.9) has only two significant figures. Therefore, your answer can only have two significant figures. When you multiply 3.45×8.9, your calculator says 30.705. But that is too many digits. You must round to two significant figures. And that gives you 31.

- Based on this, you might want to write your answer as 31×10^6. But that is not correct. In scientific notation, there should only be one number to the left of the decimal. So, you need to express your answer as 3.1×10^7. Make sure that you understand why your answer is 10^7 and not 10^5. You should think about it until it makes sense to you. Once again, think about the size of the number that you are talking about. If you make 31 smaller (by turning it into 3.1), then you must make 10^6 bigger in order to compensate.

Let's see another example:

EXERCISE 1.70.   Multiply the following numbers:

$$(3.01 \times 10^7)(4.644 \times 10^{-3}) = ???$$

Answer: Once again, just add the exponents, so you have $3.01 \times 4.644 \times 10^{(7-3)}$. The power of 10 is therefore 10^4. Now you need to multiply 3.01×4.644, and your calculator should say 13.97844. Clearly, that is too many significant figures. Your answer can only have three significant figures (because 3.01 only had three significant figures), so round up to 14.0.

So far, you have 14.0×10^4. But you need to change this number so that there is only one number in front of the decimal place. So, your answer is 1.40×10^5.

Let's get some practice. For each of the following problems, multiply the two numbers:

1.71. $(2.43 \times 10^3)(8.273 \times 10^6) = $ _____

1.72. $(7.89 \times 10^{-12})(4.6 \times 10^{-22}) = $ _____

1.73. $(4.65 \times 10^8)(5.432 \times 10^{-4}) = $ _____

1.74. $(2.59 \times 10^{-3})(2.0 \times 10^{-5}) = $ _____

1.75. $(9.4 \times 10^7)(3.45 \times 10^4) = $ _____

We just saw how to multiply two numbers that are expressed in scientific notation. Now we need to see how to divide two numbers. Recall that previously we saw that an exponent in the denominator is the same as a negative exponent:

$$\frac{1}{10^3} = 10^{-3}$$

So, instead of adding the exponents, we need to subtract. Let's see an example:

EXERCISE 1.76. Divide the following numbers:

$$\frac{3.64 \times 10^5}{4.2 \times 10^2}$$

Answer: Subtract the exponents. This gives you $\frac{3.64}{4.2} \times 10^{(5-2)}$. The power of 10 is therefore 10^3. Now divide $\frac{3.64}{4.2}$ and your calculator should say 0.866666667. Clearly, that is too many significant figures. Your answer can only have two significant figures (because 4.2 only has two significant figures), so round up to 0.87.

So far, you have 0.87×10^3. But you need to change this number so that there is one non-zero number in front of the decimal place. By changing 0.87 into 8.7, you are making a bigger number, so make 10^3 a smaller number in order to compensate. Your answer is 8.7×10^2.

Now let's get some practice dividing with scientific notation. For each of the following problems, divide the two numbers:

1.77. $(2.43 \times 10^3) / (8.273 \times 10^6) = $ _____

1.78. $(7.89 \times 10^{-12}) / (4.6 \times 10^{-22}) = $ _____

1.79. $(4.65 \times 10^8) / (5.432 \times 10^{-4}) = $ _____

1.80. $(2.59 \times 10^{-3}) / (2.0 \times 10^{-5}) = $ _____

1.81. $(9.4 \times 10^7) / (3.45 \times 10^4) = $ _____

1.3 UNITS

In the beginning of this chapter, it was said that there are two parts to every numerical answer: (1) the number, and (2) the units. So far, our discussions have focused on the number. Now we turn our attention to the units.

To express the importance of using units, let's consider the following situation. Suppose you wanted to know how long it took me to write this book. What if I told you: "12". You would probably say, "12 what?" Hours? Days? Months? What if I respond by saying "12 meters." You would wonder what I was talking about, and then you would say, "How long did it take? How much *time*?" Imagine at that point I respond by saying: "Oh, you want my answer to have units of time in it. OK. It took me exactly 12 meters *per second* to write the book." At this point, you would probably not want to talk to me anymore because there is something clearly wrong with my communication skills. I used the wrong units. A number with the wrong units is meaningless. If the answer is supposed to be in units of time, then *only* those units will make sense.

Students are often given a problem where they have to calculate something, like the mass of a compound, and they will write down an answer like this: "12". Without units, your answer is wrong. This may not seem like a big deal, but you must make it part of your habit to consider units as being *just as important* as the numerical part of your answer.

There are a few situations where we deal with unitless numbers in chemistry. For example, equilibrium constants can sometimes have no units. Whenever you come across a concept that involves a number with no units, you should be surprised, and you should try to understand why there are no units attached. For most of the calculations that you will do in this course, your answers must have units.

Now that you see why you must include the correct units in order to communicate properly, it should be pointed out that units can often be used to double check your answer. If you need to calculate a volume, and your answer comes out having units of m^2 instead of m^3, then you will have a clue that you made a mistake somewhere in your calculation. Whenever you calculate an answer to a problem, always look at your units and make sure that they make sense to you.

For any quantity that you measure, you will have a choice of acceptable units that would make sense. For example, if you were measuring the length of an object, you could record your measurement in inches, feet, miles, meters, yards, etc. All of these units correctly represent length. When measuring length, you can theoretically use any one of these units and you would be properly communicating. But having so many different types of units can be confusing for calculations, so to make things easier, the world has agreed upon certain **base units** that are the "preferred" units to use. These units are called SI units (*Le Système International d'Unités*), and they go like this:

Measure *length* in meters (m)

Measure *mass* in kilograms (kg)

Measure *time* in seconds (s)

Measure *temperature* in Kelvin (K)

Measure *amounts* in moles (mol)

There are two other SI base units, but we really don't use them in chemistry (electric current and luminous intensity), so let's just focus on the five above.

You might think that there should be more than just these five base units. What about area or volume? What about pressure? Well, it turns out that anything else you would want to measure will just be some combination of base units. Area is just $m \times m$, or m^2. Volume is just $m \times m \times m$, or m^3. Even energy is measured as a combination of the base units given above. Energy is measured with the following units:

$$\frac{kg \times m^2}{s^2}$$

Later on, we will see why energy is measured in precisely these units. This will make sense to you later. But for now, you should notice that all of the units used to measure energy (kg, m, and s) are just base units of the SI system.

We will need to see how to convert units, and we will need to take a closer look at what it means for a unit to have an exponent (like that shown above in the units for energy). But first, we need to see the prefixes that are often placed in front of the base units. Examples are *milli*gram, and *micro*second, etc. These prefixes are called *decimal multipliers*, because they involve the use of powers of 10. For example, a milligram is 10^{-3} grams. And a microsecond is 10^{-6} seconds. Instead of saying 2.4 microseconds, we could say 2.4×10^{-6} seconds, but it is faster and easier to use the term microseconds. There are many decimal multipliers, but only those commonly used by chemists are listed here:

Prefix	Power of 10	Symbol
kilo	10^3	k
deci	10^{-1}	d
centi	10^{-2}	c
milli	10^{-3}	m
micro	10^{-6}	μ
nano	10^{-9}	n

In the chart above, the last column tells you the symbol we use. A nanometer is expressed as nm. A millisecond is expressed as ms. The only symbol you might not be familiar with is μ, which is a Greek letter (*called mu*). A micrometer is expressed as μm.

To make sure that you get comfortable with the decimal multipliers above, let's do some quick and easy problems. Try to do these problems without using the chart. Take a good look at the chart for a few moments, and *then* try the problems. Express your answers using scientific notation:

1.82. 34 km ———————— m   **1.83.** 2 μs ———————— s

1.84. 482 cm ———————— m   **1.85.** 24.5 m ———————— dm

1.86. 37.2 nm ———————— m   **1.87.** 1.487 kg ———————— g

1.88. 2.46 mg ———————— g   **1.89.** 3.7 ms ———————— s

As your course moves along, you will generally see problems where the information you get is in SI units. The most common exception will be temperature. You will often be given a temperature in Celsius, and you MUST convert it to Kelvin before you can do your calculation. If you don't do the conversion, you will get the wrong answer (this is a common mistake). So, make sure you know the following conversion: whenever you are given a temperature in °C, you MUST add 273.15 to get the temperature in Kelvin. If you are told that a process takes place at 100° C, then you must convert this to 373 K.

Why didn't we convert this to 373.**15** K? Don't forget about significant figures: $100 + 273.15 = 373$. Because the number 100 does not have any decimal places, our answer cannot have any decimal places either. So, 25° C is just 298 K (this is a common temperature in problems). You may as well memorize the conversion number right now ($+273.15$), because you are going to use it a lot.

1.4 CONVERTING UNITS USING THE FACTOR-LABEL METHOD

We will soon see how to convert any units, but first we need to know how to treat units when doing calculations. Units should be treated the same way that numbers are treated. When we multiply fractions, we just multiply all of the numerators, and we multiply all of the denominators, like this:

$$\frac{3}{5} \times \frac{4}{6} = \frac{3 \times 4}{5 \times 6}$$

If there is a number that is not a fraction, then we treat it like a numerator:

$$\frac{3}{5} \times 4 = \frac{3 \times 4}{5}$$

Units are treated exactly the same way. For example:

$$kg \times \frac{m}{s} \times \frac{m}{s} = \frac{kg \times m \times m}{s \times s}$$

To make our answers shorter, we use exponents to show when we multiply more than one of the same unit. For example:

$$\frac{kg \times m \times m}{s \times s} = \frac{kg \times m^2}{s^2}$$

In the above example we see that m^2 can be used in place of $m \times m$. We did the same thing with $s \times s$. When discussing orders of magnitude earlier in this chapter, we saw that a negative exponent means it is in the denominator. The same is true here, so:

$$s^{-2} \text{ is the same as } \frac{1}{s^2}$$

Using this concept, we can rewrite the units above, so:

$$\frac{kg \times m^2}{s^2} = kg \times m^2 \times s^{-2}$$

As a side note, we do **not** use exponents when we *add or subtract* numbers with units. For example, $42\ m + 10\ m = 52\ m$ (**not** $52\ m^2$). We only use m^2 when we are *multiplying* two numbers with units of meters.

Before we move on, let's get some practice. For each of the problems below, multiply all of the units and express them using exponents, just as we did in the example above. Your answer should be written in the following type of format: $kg \times m^2 \times s^{-2}$.

1.90. $\dfrac{g}{cm \times cm \times cm} =$ **1.91.** $kg \times \dfrac{m}{s \times s} =$

1.92. $\dfrac{mol}{m \times m \times m} =$ **1.93.** $g \times \dfrac{cm \times cm}{s \times s} =$

Sometimes, you will find that you have the same units in the denominator and numerator. For example:

$$kg \times \frac{m \times m}{s \times s} \times \frac{1}{m}$$

Notice that we have an m in the numerator and an m in the denominator. Since $\dfrac{m}{m} = 1$, we can cross off one m in the numerator and one m in the denominator, like this:

$$kg \times \frac{\cancel{m} \times m}{s \times s} \times \frac{1}{\cancel{m}} = kg \times \frac{m}{s \times s}$$

After crossing off the units, we can then multiply like we did earlier. This gives us $kg \times m \times s^{-2}$.

Now let's get some practice crossing off units. For each of the problems below, cross off the units that appear in both the numerator and the denominator. Then multiply the units and express all together using exponents. Your answer should be written in the following type of format: $kg \times m^2 \times s^{-2}$.

1.94. $\dfrac{g}{mol} \times \dfrac{mol}{cm^3} =$ **1.95.** $g \times \dfrac{kg}{g} =$

1.96. $\dfrac{g}{cm^3} \times \dfrac{cm^3}{m^3} =$ **1.97.** $\dfrac{ft}{s} \times \dfrac{in.}{ft} \times \dfrac{cm}{in.} =$

1.98. $ft \times \dfrac{in.}{ft} \times \dfrac{cm}{in.} =$ **1.99.** $\dfrac{g}{mol} \times \dfrac{kg}{g} =$

Now we can explore the method that chemists use for converting units. For example, you will often have to convert grams into kilograms, or seconds into microseconds. This might seem like a simple thing to do. But as time goes on, you will encounter problems that require you to do conversions that are not so easy to just do in your head. So, let's take a close look at the method.

Our method for converting units is called the *factor-label method.* Sometimes it is referred to as *dimensional analysis,* because the method involves analyzing the dimensions (which is just a fancy term for units). This method is based on two very simple math principles:

1. Any number divided by itself is equal to one. Examples are:

$$\frac{24}{24} = 1 \qquad \frac{3.6}{3.6} = 1 \qquad \frac{½}{0.5} = 1$$

Focus carefully on the last example above. Even though we have expressed the same number in two different forms (½ and 0.5), we still get 1 as our answer. This is because ½ = 0.5, so if we divide ½ by 0.5, we just get 1. We can expand this idea to include any two measurements that represent the exact same quantity. For example, a container of 1 dozen eggs will have 12 eggs in the container. Since 1 dozen and 12 are really the same exact number, then we can say that:

$$\frac{1 \text{ dozen eggs}}{12 \text{ eggs}} = 1 \qquad \text{and} \qquad \frac{12 \text{ eggs}}{1 \text{ dozen eggs}} = 1$$

We can also use this concept with our base units to convert prefixes. For example, to show the relationship between grams and kilograms, recall that 1 kg = 1000 g. So:

$$\frac{1 \times 10^3 \text{ g}}{1 \text{ kg}} = 1 \qquad \text{and} \qquad \frac{1 \text{ kg}}{1 \times 10^3 \text{ g}} = 1$$

In each fraction above, the numerator is the exact same number as the denominator. So, each fraction is equal to 1.

We can use this concept to show the relationship between any two units that are related to each other. When comparing miles and feet, we would say that:

$$\frac{1 \text{ mile}}{5280 \text{ feet}} = 1 \qquad \text{and} \qquad \frac{5280 \text{ feet}}{1 \text{ mile}} = 1$$

That was our first simple math principle: any quantity divided by itself is just one. Now let's turn to our second math principle.

2. You can multiply any number by 1 without changing the number. For example:

$$3 \times 1 = 3$$

$$6.73 \times 1 = 6.73$$

Obviously, this is not rocket science. But if we bring together the two principles we have seen so far, we get a foolproof way to convert units. The method goes like this: we just multiply by some fraction that equals 1. Let's see an example of how this works.

Imagine that you need to measure a distance, and someone tells you that the distance is 34.7 meters. You are not so familiar with meters, so you want to have the answer expressed in feet. So, you need to convert meters into feet. You look in your textbook and you find a list that shows conversions, and you see that 1 ft = 0.3048 m.

We have seen that you can always multiply any number by 1, and we have also seen that another way to write the number 1 is like this:

$$\frac{1 \text{ ft}}{0.3048 \text{ m}} \quad \text{or} \quad \frac{0.3048 \text{ m}}{1 \text{ ft}}$$

So, we can multiply our starting number (34.7 meters) by either one of the fractions above without changing the number. Since we want to convert meters into feet, we will need to choose the fraction that has meters in the denominator—that is the only way that units of meters will be crossed off completely:

$$34.7 \text{ m} \times \frac{1 \text{ ft}}{0.3048 \text{ m}}$$

We have not changed our number at all, because we have just multiplied by one. Now, we can cross off units, because we have meters in the numerator and meters in the denominator:

$$34.7 \ \cancel{m} \times \frac{1 \text{ ft}}{0.3048 \ \cancel{m}} = \frac{34.7 \times 1 \text{ ft}}{0.3048}$$

Now we have our answer in feet. All you have to do is divide 34.7 by 0.3048, and then report your answer with the appropriate significant figures. In this case, we should report three significant figures, so our answer is 144 ft. Notice that our answer is expressed with the appropriate significant figures *and* with the appropriate units.

You can use this method to convert any two units for which you know the conversion factor (like we did above, when we knew the conversion factor 1 ft = 0.3048 m). Anytime you do a conversion using this method, you will have *two* choices for which multiplier to use. In the example above, we could have used

$$\frac{1 \text{ ft}}{0.3048 \text{ m}} \quad \textit{or} \quad \frac{0.3048 \text{ m}}{1 \text{ ft}}$$

but only the first fraction gave us a meaningful answer. If we had chosen the second fraction, we would have gotten an answer with too many units:

$$34.7 \text{ m} \times \frac{0.3048 \text{ m}}{1 \text{ ft}} = 34.7 \times 0.3048 \times \text{m}^2 \times \text{ft}^{-1}$$

If you look at the units you would get in the above equation, you immediately realize that you chose the wrong multiplier. You only use the second fraction if you want to change from feet to meters instead of meters to feet. After you do a few practice problems, you should get comfortable with quickly choosing which of the two multipliers to use.

Let's start off with some simple problems, and then we'll build up the complexity as we go along. As you start out, you will probably wonder why you need to use this method at all. You will probably say that you could have just gotten the answer by doing the calculation in your head. But as we build the complexity of the problems, you will soon see that it is not always so easy to do the calculations in your head. By using the method given above, you will see that even these more complex problems should still be straightforward to do.

Let's begin with some simple conversions:

EXERCISE 1.100. An object weighs 24.7 lb. What is its mass in kg? Use the following information:

$$1 \text{ lb} = 0.4536 \text{ kg}$$

Answer: You are given the conversion, so you must use one of the following two fractions:

$$\frac{1 \text{ lb}}{0.4536 \text{ kg}} = 1 \quad \text{and} \quad \frac{0.4536 \text{ kg}}{1 \text{ lb}} = 1$$

Since you are converting pounds to kilograms, use the second fraction, so that pounds will cross off and you will be left with units of kilograms:

$$24.7 \text{ lb} \times \frac{0.4536 \text{ kg}}{1 \text{ lb}} = 11.20392 \text{ kg}$$

Now, round your answer so that you display the appropriate significant figures. 24.7 has three significant figures, so your answer must also have three significant figures: **11.2 kg**

PROBLEM 1.101. An object weighs 153.2 lb. What is its mass in kg?

Answer: _____

PROBLEM 1.102. An object has a mass of 2.0 kg. What is the weight of the object in pounds?

Answer: _____

PROBLEM 1.103. An object is 54.6 inches tall. Convert the height into centimeters, using the following conversion: 1 in. = 2.54 cm.

Answer: _____

1.5 USING MORE THAN ONE CONVERSION FACTOR

Sometimes, we need to use two or more conversion factors in order to convert our answer into the correct units. Let's see an example:

EXERCISE 1.104. The distance between two points is 1.07 miles. Convert this distance into inches using the following information:

$$1 \text{ mile} = 5280 \text{ feet} \qquad 1 \text{ foot} = 12 \text{ inches}$$

Answer: In order to convert miles into inches, you will have to convert miles into feet, and then feet into inches. *But you don't have to do this in two steps.* You can do it all in one step:

$$1.07 \text{ miles} \times \frac{5280 \text{ ft}}{1 \text{ mile}} \times \frac{12 \text{ in.}}{1 \text{ ft}} = 6779.52 \text{ in.}$$

Finally, round your answer to three significant figures, so your answer is: **6780 in.**

The trick here is that you don't have to do the problem in two steps. It is quicker and easier to do everything in one step, using both conversion factors. If you do it right, everything will cross off to give you the units you want.

This becomes important as we move on to problems with *more* than two conversion factors. Sometimes, you will do problems with five or six conversion factors. If you try to do one at a time, it takes too long. So, you need to get accustomed to using multiple conversion factors at the same time. Let's try another problem like Exercise 1.104. Make sure to do your conversions all in one step.

PROBLEM 1.105. The distance between two points is 1.07 miles. Convert this distance into yards using the following information: 1 mile = 1.609 km, and 1 km = 1094 yards.

Answer: _____

Sometimes, we have to convert units when exponents are involved. Let's see an example:

EXERCISE 1.106. The volume of an object is 34.0 ft^3 (cubic feet). Convert this into m^3 (cubic meters) using the following information: 1 ft = 0.3048 m.

Answer: Think about what it means when you have an exponent on a unit. ft^3 means ft $\times$ ft $\times$ ft. This should make sense, because the volume of an object is just (width) $\times$ (height) $\times$ (length). In order for all of the units of feet to cross off so that you are left with units of meters, you will need to use your conversion factor *three times*:

$$34.0 \text{ ft} \times \text{ft} \times \text{ft} \times \frac{0.3048 \text{ m}}{1 \text{ ft}} \times \frac{0.3048 \text{ m}}{1 \text{ ft}} \times \frac{0.3048 \text{ m}}{1 \text{ ft}}$$

When you calculate your answer and round to three significant figures, you get 0.963 m^3.

PROBLEM 1.107.   The volume of an object is 12 cubic meters. Convert this into cubic centimeters.

Answer: _____

In the previous problems, you had to change only one set of units (such as meters to feet). Now, let's consider some problems where you have to change units in the numerator AND in the denominator. Let's see an example:

EXERCISE 1.108.   A molecule is traveling at 520 m/s. Convert this speed into miles/hour.

Answer: In this example, you will need to convert the numerator from meters into miles:

$$520 \ \frac{\text{m}}{\text{s}} \times \frac{1 \text{ km}}{1000 \text{ m}} \times \frac{1 \text{ mile}}{1.609 \text{ km}}$$

And you will also need to convert the denominator from seconds into hours, using the multipliers:

$$\frac{60 \text{ s}}{1 \text{ min}} \times \frac{60 \text{ min}}{1 \text{ h}}$$

As we said before, always try to do everything in one step. There are four conversion factors here, so you should multiply everything at the same time. When you do this, all of the units cross off except miles/hour:

$$520 \ \frac{\text{m}}{\text{s}} \times \frac{1 \text{ km}}{1000 \text{ m}} \times \frac{1 \text{ mile}}{1.609 \text{ km}} \times \frac{60 \text{ s}}{1 \text{ min}} \times \frac{60 \text{ min}}{1 \text{ h}}$$

Round your answer to two significant figures, and you get: 1200 miles/h. Here is a situation where you should use scientific notation in order to be clear about how many significant figures you are talking about. When you made the calculation above, your calculator said 1163.45556, but you had to round up to two significant figures, to 1200. To make it clear that we are talking about two significant figures (and not the exact number 1200, with four significant figures), use scientific notation: 1.2×10^3 miles/h. This makes it clear that there are two significant figures.

PROBLEM 1.109. An object is traveling at 342.3 meters per second. Convert this speed into miles per hour.

Answer: _____

PROBLEM 1.110. An object is traveling at 22 miles per hour. Convert this speed into kilometers per minute (1 mile = 1.609 km).

Answer: _____

1.6 SPECIAL CONVERSION FACTORS

In the previous section, we saw how to convert units. In all of our examples, we were converting either length into length, or time into time, etc. It makes sense to convert miles into inches, or to convert seconds into years. But does it make any sense to convert *miles* into *years*? The units of miles and the units of years are measuring two completely different properties: length and time. It turns out that physicists do in fact convert these units. Let's see how.

Length and time are actually intimately related to each other through the special theory of relativity. Now, we don't want to get into relativity here, because this is not a physics course, but on a superficial level, we can see the relationship between length and time if we just consider measurements that we regularly make. When we measure velocity, we are measuring the *distance* traveled per unit of *time*. Now, here comes the trick: if you have something that travels at constant velocity, such as light, then you can measure *how far* the light travels *in terms of the time* it takes to travel that far. When physicists talk about a "light year", they are not referring to a unit of time. Rather, they are referring to a distance—the distance that light travels in one year. When they say that a galaxy is one million *light years* away from us, they are talking about its *distance* from us.

In chemistry we also have terms that relate one set of units to another set of units. There are a few of these, and if you understand them, then you will have unlocked the keys to solving many problems. Since they are so important for solving problems, we will focus on them throughout this book. In this section, we will focus on one such special conversion factor: *density*.

The density of a substance measures the *mass* that is contained in a certain *volume*. The units are usually in g/cm^3. Another way of saying cm^3 is milliliter (mL). So, the units of density can also be expressed as g/mL. Before we can see how to use density as a special conversion factor, let's first get some practice calculating densities.

EXERCISE 1.111. A substance has a mass of 1.62×10^3 mg, and is contained in a cube with the following dimensions: 21.3 mm $\times$ 10.7 mm, $\times$ 12.0 mm. Calculate the density in g/cm^3.

Answer: We are asked to calculate the density, and we are given the mass and the volume. So, we just need to divide the mess by the volume. That gives us:

$$\frac{\text{mass}}{\text{volume}} = \frac{1.62 \times 10^3 \text{ mg}}{21.3 \text{ mm} \times 10.7 \text{ mm} \times 12.0 \text{ mm}}$$

BUT, we were asked to give our answer in units of g/cm^3. So, we must convert mg into g, and we must convert mm into cm. To convert mg into g, we use the following multiplier:

$$\frac{1 \text{ g}}{1000 \text{ mg}}$$

and to convert mm into cm, we will use the following:

$$\frac{10 \text{ mm}}{1 \text{ cm}}$$

Now, we can solve for our answer:

$$\frac{1.62 \times 10^3 \text{ mg}}{21.3 \text{ mm} \times 10.7 \text{ mm} \times 12.0 \text{ mm}} \times \frac{1 \text{ g}}{1000 \text{ mg}} \times \frac{10 \text{ mm}}{1 \text{ cm}} \times \frac{10 \text{ mm}}{1 \text{ cm}} \times \frac{10 \text{ mm}}{1 \text{ cm}}$$

$$= 0.592 \text{ g/cm}^3$$

Notice that the answer was recorded with three significant figures.

PROBLEM 1.112.   A substance has a mass of 0.012 kg, and is contained in a volume with the following dimensions: 25.4 mm × 12.3 mm, × 13.4 mm. Calculate the density in g/cm^3 (and don't forget about significant figures).

Answer: _____

Now that we have seen how to calculate the density of a substance, we are ready to see how it can be used as a special conversion factor. Density tells us how much mass can be found in a particular volume. So, we can use this to convert from mass to volume, and vice versa. Let's say we know the density of a substance at a certain temperature, then we should be able to calculate the volume if we are given the mass. Similarly, we should be able to calculate the mass if we are given the volume. We do this in exactly the same way that we converted units in the previous section of this chapter. Let's see an example of how this works:

EXERCISE 1.113.   At 25° C, the density of water is 1.00 g/mL. What is the mass in grams of 1.78 gallons of water? (1 gallon = 3785 mL).

Answer: Before you get started, note that you will need to convert gallons into milliliters, because your final answer must be expressed in terms of milliliters. So, you need to use the conversion given above, which is:

$$1.78 \text{ gallons} \times \frac{3785 \text{ mL}}{1 \text{ gallon}}$$

Now you are ready to do your calculation. You are given the density of water at room temperature. So, you know the relationship between mass and volume. You can treat this in exactly the same way that you did when you were converting units in the previous section. The density of water tells you that 1.00 g = 1 mL. Therefore, you can set up two fractions that are both equal to 1, just as you did before:

$$\frac{1.00 \text{ g}}{1 \text{ mL}} = 1 \qquad \text{and} \qquad \frac{1 \text{ mL}}{1.00 \text{ g}} = 1$$

In the fractions above, 1 mL is treated as an exact number rather than a number with one significant figure because we define density as the amount of grams in exactly 1 mL of volume. So, you can treat the number 1 mL as having an infinite amount of significant figures.

Now you just need to multiply the volume (that was converted to milliliters) by one of the fractions above. Since you want the units of volume to cancel off, use the first fraction:

$$1.78 \text{ gallons} \times \frac{3785 \text{ mL}}{1 \text{ gallon}} \times \frac{1.00 \text{ g}}{1 \text{ mL}} = 6740 \text{ g}$$

Notice that you record your answer with three significant figures, which makes the answer 6740 g. In this example, you have to use scientific notation to eliminate any ambiguity regarding how many significant figures you have. The answer is expressed as: **6.74×10^3 g**

Before we conclude the chapter, there are two subtle points that should be mentioned. When we use density as a conversion factor, we need to be careful of two things. There are two major differences between the conversions we saw in the previous section and the conversion we are doing here with density:

1. We need to consider the role of significant figures. Unlike the regular conversion factors, this special conversion factor (density) must be reported with the correct number of significant figures. To see what we mean by this, let's consider a regular conversion factor. When we say that 1 kg = 1000 g, we are really saying that

$$1.00000000 \ldots \text{ kg} = 1000.00000 \ldots \text{ g}$$

In other words, when we talk about this conversion (1 kg = 1000 g), we do not mean to imply that the term 1 kg has only one significant figure. It actually has an infinite number of significant figures. But, when we say that the density of an object is 0.592 g/cm^3, we are saying that we only know the value of this conversion factor to three significant figures. Whenever you deal with density (whether you are calculating a density or using it as a special conversion factor), you should take special care to follow the rules of significant figures.

2. When we deal with regular conversion factors, we are dealing with fundamental relationships that are always true. For example, consider this regular

conversion factor: 1 kg = 1000 g. This relationship is not dependent on any-thing else. It is true under any conditions. But when we say that 1.00 g = 1 mL, we are talking about a relationship that is dependent on many factors, such as temperature and pressure. Therefore, this relationship only holds at a specific temperature and pressure. If we change the temperature and pressure, then the density will change as well.

Let's do two final problems that incorporate everything we have seen in this chapter.

PROBLEM 1.114.   At a certain temperature, the density of water is 0.9978 g/mL. What is the mass in grams of 1.782 gallons of water at this temperature? (1 gal-lon = 3785 mL). Report your answer using scientific notation, and make sure to use appropriate significant figures.

Answer: _____

PROBLEM 1.115.   At 20° C, the density of water is 0.9978 g/mL. What volume will be occupied by 1.75 kg of water at this temperature? Report your answer us-ing scientific notation, and make sure to use appropriate significant figures.

Answer: _____

CHAPTER 2

COUNTING ATOMS AND MOLECULES

In this chapter, we will use the tools and techniques that we learned in the previous chapter, and we will begin to solve problems that involve the relationship between atoms and molecules. The first few sections provide the basic concepts and terminology that you will use in the rest of the chapter. The remaining sections are dedicated to common types of problems, representing the fundamental calculations that you will need to master (not only for your exams, but also to serve as a foundation for material that we will learn in subsequent chapters). The problems and techniques will slowly build in complexity, but the methods for solving these problems are not difficult. Once you get practice, you should find that these problems are all very straightforward.

2.1 EMPIRICAL AND MOLECULAR FORMULAS

We will now develop an analogy that we will use again and again throughout this book. It will help you understand what you are doing when you perform the calculations necessary to solve the problems in the rest of this chapter. In our analogy, we will think of atoms as Lego pieces. At first, this analogy might seem simple and unnecessary. But as you begin to deal with more challenging problems, the Lego analogy will help you to have a better understanding of the principles involved in solving problems. In the next chapter, when we discuss stoichiometry, we will rely heavily on the use of our Lego analogy.

Lego pieces come in different sizes, shapes, and colors. The number of notches on the top of a Lego piece tells us how many other Lego pieces can connect with it to build a structure. We can imagine building very simple structures with our Lego pieces, for example:

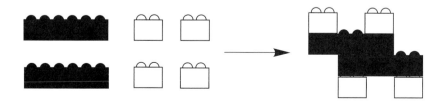

Let's say we wanted to take some Lego pieces and actually build the structure shown above. The figure above tells us that we would need two black pieces and four white pieces. If we wanted to build 100 structures like the one above, we would need 200 black pieces and 400 white pieces. The more pieces we have, the more of these structures we can build, but we need to have the pieces in a certain ratio. Specifically, we need to have one black piece for every two white pieces. We can express this in a simple way: $(Black)_1(White)_2$. This expression does not tell us how the pieces are connected. In fact, it does not even tell us how many Lego pieces we need to build one structure. If I gave you one black piece and two white pieces, you would not be able to build even one structure.

Instead, we can choose an expression that *does* tell us how many pieces you need to build one structure. In our example, we would express it like this: $(Black)_2(White)_4$. This expression is similar to the one before, but now it tells the information we need to build one structure.

In our analogy, the individual Lego pieces represent atoms, and the structure represents a molecule:

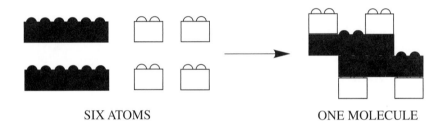

SIX ATOMS   ONE MOLECULE

So atoms are the building blocks of molecules, in the same way that Lego pieces are the building blocks of Lego structures.

We saw two ways to express the recipe for building one structure: $(Black)_2$ $(White)_4$ tells us the number of each type of piece that is needed to build one structure. And $(Black)_1(White)_2$ just tells us the relative ratio of white and black pieces. The same is true when we use atoms to build molecules. We can express the molecules in a way that tells us exactly how many atoms are needed to build one molecule, or we can just show the ratio of atoms needed. A common example is N_2O_4 vs. NO_2. The first formula tells us the molecule is made out of two nitrogen atoms and four oxygen atoms. The second formula only tells us the ratio of nitrogen to oxygen atoms, but does *not* tell us how many atoms are actually needed to build the molecule. The first formula (N_2O_4) is called the *molecular formula*, and the second formula (NO_2) is called the *empirical formula*. Notice that we do not use a subscript next to the nitrogen in NO_2. When there is no subscript, we assume that it means 1.

In some cases, the molecular formula *is* the empirical formula. For example, consider a water molecule, which is made with two hydrogen atoms and one oxygen atom. The molecular formula is therefore H_2O, and the empirical formula is just H_2O.

EXERCISE 2.1. Consider the following molecular formula:

$$C_4H_{10}$$

Is this also an empirical formula? If not, then write out the empirical formula.

Answer: To determine if this is an empirical formula, you need to see if the smallest whole numbers are used. In this case, you could use smaller whole numbers to express the ratio of atoms in the molecule. Both 4 and 10 can be divided by 2, giving C_2H_5. Therefore, the empirical formula is C_2H_5. Notice that an empirical formula uses the smallest *whole* numbers that express the ratio. So, your answer is not $CH_{2.5}$ even though that expresses the correct ratio.

For each of the following molecular formulas below, write the empirical formula next to it. In some cases, the empirical formula might be the same as the molecular formula.

2.2. C_2H_6 _____   **2.3.** C_3H_8 _____

2.4. CO_2 _____   **2.5.** H_2SO_4 _____

2.6. $C_6H_{12}O_6$ _____   **2.7.** Hg_2I_2 _____

2.8. Al_2O_3 _____   **2.9.** C_5H_5N _____

Empirical formulas are mostly an historical artifact. In the old days, chemists determined the formulas of compounds through combustion analysis. This technique allowed for chemists to determine empirical formulas, rather than molecular formulas. But in applying modern-day technology, we have all kinds of fancy techniques (NMR, Mass Spec, X-Ray Diffraction, etc.) that allow us to easily determine the molecular formula. It is somewhat of a mystery why textbooks continue to teach empirical formulas. For whatever reason, empirical formulas have not yet left the textbooks, so you will need to know what they are. But for the most part, you will not use empirical formulas as you move along in this course. The focus will be on molecular formulas.

In the rest of this chapter, most of the problems will start by giving a molecular formula. You will soon see that the molecular formula is a very valuable piece of information, because you can usually extract the information you need from the molecular formula. It is a common mistake for students to miss this point. When a problem says: "22.4 g of Al_2O_3 . . ." you have actually been given *two* pieces of information: 22.4 g *and* Al_2O_3. If you fail to see the second piece of information, you will not have enough information to solve the problem. We will see many examples of this soon. But first, we have to go over a few more basics.

2.2 MOLECULES AND MOLES

Let's revisit our Lego analogy, in an effort to keep things as simple as possible. Imagine that you need to build a lot of identical Lego structures, like the structure we saw in the previous section:

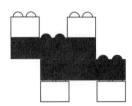

Your task is to build as many of these structures as possible. So you call a Lego distribution warehouse, and you ask if you can purchase Lego pieces in bulk. The salesperson tells you that they only sell Lego pieces by the barrel. Each barrel contains over 6 million pieces. In fact, all barrels come with exactly 6,022,042 Lego pieces. It is a strange number, but that's the way they are sold.

Then the salesperson tells you some good news. In addition to selling barrels of individual Lego pieces, the warehouse also sells barrels of ready-made structures. They have a large variety of ready-made structures, and it turns out that they sell barrels of the structure that you want to build—$(Black)_2(White)_4$. When you ask how many structures are in each barrel, you get the same answer you got before: 6,022,042. But this time, we are not talking about individual pieces. This time, we are talking about over 6 million *preassembled structures*. If we were to dump out the contents of one of these barrels, and then take apart all of the structures, we would have over 36 million individual Lego pieces from one barrel alone.

We said before, that our analogy uses individual Lego pieces to represent atoms, and Lego structures to represent molecules. Now we have added a new term: a barrel, which represents the term "mole". A mole is just a large number (a barrel) of molecules. In our Lego analogy, each barrel had 6,022,042 (or 6.022042×10^6) Lego structures. Well, it turns out that a mole of molecules is a lot larger than a barrel of Lego structures. The number is actually 17 orders of magnitude larger (100,000,000,000,000,000 times larger). So, there are 6.022042×10^{23} molecules in a mole. Since most of our calculations will involve other numbers that have fewer significant figures, let's make things simpler, and let's use fewer significant figures. From now on, we will use the number **6.022** $\times 10^{23}$.

The "mole" is a very important term. It is the SI unit that is used to measure the number of molecules in a substance. For example, when we measure the number of molecules, we report our answer as 3.72 mol, or 5.3 mol, etc. As we have already discovered, one mole is defined as 6.022042×10^{23} molecules. This number is also called *Avogadro's number*. It is clearly a *VERY* large number. The term "mole" is the term that we use to connect our macroscopic world to the world of very small atoms and molecules.

EXERCISE 2.10.   Count the number of oxygen atoms in 3.45 moles of H_2SO_4.

Answer: This might seem like simple math, and you might be tempted to do this in your head. But you need to get into the habit of always using the factor-label method to solve the problem. You will see many more problems that are not so easy to do in your head, so you need to get practice using the factor-label method now. In addition, the factor-label method serves as a way to double-check your answer. If you divide numbers instead of multiplying them (or vice versa), then the factor-label method will let you know you did it wrong, because the units won't cross off properly in the end.

In order to use the factor-label method, use the following information:

$$1 \text{ mole of } H_2SO_4 = 6.022 \times 10^{23} \text{ molecules of } H_2SO_4$$

and

$$1 \text{ molecule of } H_2SO_4 = 4 \text{ oxygen atoms}$$

Using this information, you can make the following conversion:

$$3.45 \text{ mol} \times \frac{6.022 \times 10^{23} \text{ molecules}}{1 \text{ mol}} \times \frac{4 \text{ oxygen atoms}}{1 \text{ molecule}}$$

Notice that two conversion factors were used here. The first one converts moles into a number of molecules, and the second one converts a number of molecules into a number of oxygen atoms.

When you use a calculator to multiply the numbers above (except for 10^{23}), your calculator says 83.1036. So, you have 83.1036×10^{23}. Before you write down your final answer, you need to make sure that your numbers and units are expressed properly. There are only three significant figures in 3.45 moles, so your answer must only have three significant figures. That gives you 83.1×10^{23}. Next, you need to modify the power of 10, so that there will only be one digit in front of the decimal place. By moving the decimal over one place to the left, you are making the number smaller, so you must compensate by making the power of 10 larger. This gives 8.31×10^{24}. Finally, you need to make sure your answer expresses the units properly:

$$\textbf{8.31} \times \textbf{10}^{\textbf{24}} \textbf{ oxygen atoms}$$

PROBLEM 2.11.   Count the number of hydrogen atoms in 0.034 moles of propane, C_3H_8.

Answer: _____

PROBLEM 2.12.   Count the number of oxygen atoms in 0.872 moles of SO_3.

Answer: _____

PROBLEM 2.13.   Count the number of boron atoms in 2.7 moles of B_2H_6.

Answer: _____

2.3   ATOMIC MASS

In this section, we will focus on how we measure the mass of a single atom or a mole of atoms.

Let's start with a single atom. If you try to measure the mass of a single hydrogen atom in terms of *grams*, you will get a *very* small number (with an order of magnitude of 10^{-24} g). So, rather than using very small numbers all of the time, chemists created a new term to express the mass of individual atoms, and even individual molecules. This term is called an ***atomic mass unit***, and it is defined as 1 gram divided by Avogadro's number (6.02×10^{23}). So, one atomic mass unit equals $1.66053873 \times 10^{-24}$ g.*

You might wonder why the chemists didn't just define the new unit as being exactly 1.0×10^{-24} g, and there is a good answer. By defining an atomic mass unit as 1 gram divided by Avogadro's number, chemists have actually made things much easier for us. To see why, we need to take a short tour of the periodic table.

If you look at the periodic table in your textbook (or any other periodic table), you will find a number in the bottom of every box:

For example, the number at the bottom of the box for hydrogen is 1.008. That number tells us that one hydrogen atom has a mass of 1.008 atomic mass units, AND it tells us that one mole of hydrogen atoms is 1.008 grams. For purposes of calculations, everything is simplified because this one number (1.008) refers to the mass of an individual atom (measured in atomic mass units), ***and*** the mass of a mole (measured in grams). This only works because of the way chemists chose to define the atomic mass unit. If they had defined an atomic mass unit as 1.0×10^{-24} grams, then the mass of one hydrogen atom would have a different numerical value than the mass of a mole of hydrogen atoms.

The accepted symbol for atomic mass units is "u". So, a hydrogen atom has a mass of 1.008 u. A carbon atom has a mass 12.01 u.

In this way, we can calculate the mass of a single *molecule* by adding up the mass of each atom in the molecule. For example, a water molecule is composed of two hydrogen atoms and one oxygen atom (H_2O). The mass of a hydrogen atom is 1.008 u, and the mass of an oxygen atom is 16.00 u. Therefore, the mass of a water molecule is just $2(1.008 \text{ u}) + (16.00 \text{ u}) = 18.02$ u.

---

*To be more accurate, chemists actually defined the atomic mass unit based on the mass of a carbon-12 atom (the most abundant isotope of carbon) divided by 12. Another way of saying this is to simply state that an atomic mass unit is just 1 gram divided by Avogadro's number.

It is unlikely that your instructor will expect you to memorize the mass of every element on the periodic table, but you should absolutely know how to use the periodic table to find the information you need. In general, there are several elements that you will find yourself using again and again, so you may as well memorize these now. Hydrogen is 1.008 (or 1.0079, depending on how many significant figures your periodic table shows), carbon is 12.01, and oxygen is 16.00. It is OK if you don't memorize these now, but you might find that memorizing them could save you some time when you are doing problems.

EXERCISE 2.14.   What is the mass of one molecule of carbonic acid (H_2CO_3)?

Answer: This molecule is composed of two hydrogen atoms, one carbon atom, and three oxygen atoms. Using a periodic table, we see that the mass of one molecule of carbonic acid is just:

$$2(1.008 \text{ u}) + (12.01 \text{ u}) + 3(16.00 \text{ u}) = 62.03 \text{ u}$$

PROBLEM 2.15.   What is the mass of one molecule of barium chromate ($BaCrO_4$)?

Answer (consult the periodic table in your textbook): _____

PROBLEM 2.16.   What is the mass of one molecule of zinc oxalate (ZnC_2O_4)?

Answer: _____

PROBLEM 2.17.   What is the mass of one molecule of silver bromide (AgBr)?

Answer: _____

2.4 MOLAR MASS

In the previous section we saw how to calculate the mass of a single molecule by adding up the mass of each of the atoms in the molecule. Our answer was expressed in atomic mass units. If we wanted to measure the mass of a *mole*, we would use the same exact numbers, but our units would change from atomic mass units to *grams*. If every molecule of water has a mass of 18.02 u, then every mole of water will have a mass of 18.02 grams. Notice how elegant that is. Once again, that's the beauty of the way chemists chose to define an atomic mass unit.

For any compound, one mole means 6.022×10^{23} molecules (by definition). But the mass of those molecules will be different from one compound to the next. For example, a mole of water molecules is 6.022×10^{23} molecules, and it weighs 18.02 g. A mole of liquid hydrogen is 6.022×10^{23} H_2 molecules, and it weighs only 2.016 g. In each case, we are dealing with the same number of molecules. But in the case of water, each molecule is made up of two hydrogen atoms *and* one oxygen atom; whereas in the case of H_2, each molecule is made up of only two hydrogen atoms.

So, we see that a mole of one substance will not have the same mass as a mole of another substance. For any substance, we can determine how much mass will be in one mole of that substance. And we can say it like this: one mole of water molecules will have a mass of 18.02 grams. But there is a simpler way to say this. We can say that the ***molar mass*** is 18.02 ***g/mol***. Notice the units that we are using now. These units (***g/mol***) indicate that we are talking about the amount of mass *per mole* of the substance. Before moving on, let's make sure that we are clear on the difference between the two terms that appear in these units (***g*** and ***mol***). To see the difference between these measurements, imagine that you have a room filled with people. You can add up their collective mass, or you can count the number of people. Those are clearly two different measurements.

To calculate the molar mass of a compound, all you need is the molecular formula. Let's see an example of how this works.

EXERCISE 2.18. Calculate the molar mass of ethylene, which has the following molecular formula: C_2H_4.

Answer: You are given the molecular formula: C_2H_4. That means that a *molecule* of ethylene consists of two carbon atoms and four hydrogen atoms. In the previous section you saw that one *molecule* of ethylene should have the same mass as (2 carbon atoms) + (4 hydrogen atoms). Similarly, one *mole* of ethylene should have the same mass as (2 *moles* of carbon atoms) + (4 *moles* of hydrogen atoms). All you need to know is the mass of a mole of carbon atoms and the mass of a mole of hydrogen atoms. And you can find that information on the periodic table, which tells you that carbon atoms have a mass of 12.01 g/mol, and hydrogen atoms have a mass of 1.008 g/mol.

With this information, you can calculate the molar mass of ethylene. C_2H_4 should equal 2(12.01 g/mol) + 4(1.008 g/mol) = 28.05 g/mol (don't forget about significant figures—when you do the math above, your calculator will say 28.052, but your answer can only have *two* decimal places because 12.01 only had two decimal places). So, the molar mass of ethylene is 28.05 g/mol (if you had used a periodic table that gave you more significant figures, then your answer would have had more significant figures in it). Notice the units of your answer: g/mol. This should make sense because your answer describes the amount of grams in one mole of ethylene.

In just a few moments, you will see how to use the molar mass as a special conversion factor to convert moles into mass. But first, let's get some practice calculating molar masses using the periodic table. This might seem trivial, but this skill will be ***absolutely essential*** in the rest of the problems of this chapter, so you must make sure that you have mastered this skill right now. You will rarely be given the molar mass of a compound—instead, you will have to calculate it yourself using the molecular formula.

Using the periodic table in your textbook, calculate the molar mass of the following compounds:

2.19. C_2H_6 _____   **2.20.** C_3H_8 _____

2.21. CO_2 _____   **2.22.** H_2SO_4 _____

2.23. $C_6H_{12}O_6$ _____   **2.24.** Hg_2I_2 _____

2.25. Al_2O_3 _____   **2.26.** C_5H_5N _____

In this section, we have seen how to calculate the molar mass of a compound, simply by using the periodic table. It is important to realize that chemists have measured the atomic mass of each atom, and these **_known values_** contain MANY significant figures. The periodic tables shown in most textbooks will give atomic masses measured with only four or five significant digits. For example, hydrogen will be listed as 1.008 amu, or 1.0079 amu. We can use these values to calculate the molar mass of any compound, and our answer will have either four or five significant figures. In most problems, you will never need to use a molar mass with more than four or five significant figures. After all, most problems give you data that are reported with either three or four significant figures. But, you should realize, that atomic mass values are known with more significant figures. Therefore, if it is necessary, you should be able to calculate the molar mass of a compound using more than five significant figures (assuming you are given atomic mass information that is reported with more than five significant figures).

For example, some textbooks will report atomic masses with **_six_** significant figures. Hydrogen will be reported as 1.00794 amu and oxygen will be reported as 15.9994 amu. If we use these numbers to calculate the molar mass of H_2O, we will get:

$$2(1.00794 \text{ g/mol}) + (15.9994 \text{ g/mol}) = \textbf{\textit{18.0153 g/mol}}$$

So, we see that the molar mass of any compound is a value that is known to many significant figures. In general, we will only use four or five significant figures when solving problems.

2.5 MOLAR MASS AS A CONVERSION FACTOR: INTERCONVERTING MOLES AND MASS

In Chapter 1, we learned how to convert units using the factor-label method, and we saw an example of a special conversion factor that helped us convert one type of measurement into another. Specifically, we saw how to use the *density* of a substance to convert from units of mass *into* units of volume, and vice versa. Although mass and volume are measured with completely different units, we were able to convert them using density as a conversion factor.

In this section, we will use a similar technique; a similar type of special conversion factor. We will convert mass into moles, and vice versa, using molar mass as a conversion factor. But this time, we don't need to be given the conversion factor—we can calculate it ourselves. When we talked about density as a special conversion factor, we saw problems where we were given the density at a certain temperature, and then we made our calculations. If we didn't have the density, then we wouldn't be able to use it as a conversion factor. But here, when we are talking about converting mass and moles, we can calculate the conversion ourselves if we are given the molecular formula. All we need is access to a periodic table, and we will be able to calculate the molar mass. And that is why the molecular formula always provides us with an important piece of information when solving problems.

Now let's see how we can use the molar mass of a compound as a special conversion factor. Let's revisit the example of ethylene again. Previously, we saw how to use the molecular formula of ethylene (C_2H_4) to calculate its molar mass. We calculated that the molar mass of ethylene is 28.05 g/mol. This means that:

$$1 \text{ mole of ethylene} = 28.05 \text{ gram of ethylene}$$

We can use this to convert moles into grams, and vice versa, using the following conversion factors:

$$\frac{28.05 \text{ g}}{1 \text{ mol}} = 1 \qquad \text{and} \qquad \frac{1 \text{ mol}}{28.05 \text{ g}} = 1$$

This is the same kind of conversion we saw in Chapter 1 (the factor-label method for converting units). But be careful: the relationships shown above are for ethylene. Different compounds will have different molar masses.

Let's see an example of how we can use these conversion factors:

EXERCISE 2.27.   Calculate the mass (in grams) of 2.72 moles of CO_2.

Answer: This problem is asking you to convert moles into mass. To do this, you will need the molar mass, which the problem has not directly given. But you do have the molecular formula, and that is all you need to calculate the molar mass.

The molecular formula is CO_2. So, one *mole* should have the same mass as (1 *mole* of carbon atoms) + (2 *moles* of oxygen atoms). All you need to know is the mass of a mole of carbon atoms and the mass of a mole of oxygen atoms. By using the periodic table you'll get: 1 mole of CO_2 should equal (12.01 g) + 2(16.00 g) = 44.01 g (don't forget about significant figures). So, the molar mass of CO_2 is 44.01 g/mol.

Now you have the formula you need to solve the problem:

$$1 \text{ mol of } CO_2 = 44.01 \text{ g}$$

So, set up the following two conversion factors:

$$\frac{44.01 \text{ g}}{1 \text{ mol}} = 1 \qquad \text{and} \qquad \frac{1 \text{ mol}}{44.01 \text{ g}} = 1$$

You just have to decide which one of these to use. In this problem, you are converting moles into grams, so use the first conversion factor:

$$2.72 \text{ mol CO}_2 \times \frac{44.01 \text{ g}}{1 \text{ mol CO}_2} = 120 \text{ g}$$

$$= 1.20 \times 10^2 \text{ g}$$

Notice that your answer is expressed with three significant figures.

PROBLEM 2.28. Calculate the mass (in grams) of 0.89 moles of $C_6H_{12}O_6$. Express your answer using scientific notation.

PROBLEM 2.29. You have a water bottle filled with 67.2 grams of water (H_2O). Calculate how many moles of water in are in the water bottle. (Hint: after calculating the molar mass of H_2O, write out the two conversion factors, and carefully consider which one you would use. In this problem you are converting from grams to moles instead of moles to grams).

PROBLEM 2.30. You have a barrel filled with 25.6 kg of water (H_2O). Calculate how many moles of water in are in the barrel. (This problem is the same as the previous one, only you are starting with kg instead of g, so you will need to use an additional conversion factor in your calculation—to convert kg into g).

PROBLEM 2.31. A piece of copper has a mass of 2.01 kg. Calculate the number of moles of copper atoms. [In this problem, there is no molecular formula, because you are dealing with individual copper atoms, which simplifies the problem because you don't need to first *calculate* the molar mass. You can get the molar mass directly from the periodic table by looking at copper (Cu) on the periodic table].

PROBLEM 2.32. Calculate the mass (in grams) of 7.04 moles of $CaCO_3$. Express your answer using scientific notation.

2.6 USING MOLAR MASS AS A CONVERSION FACTOR TWO TIMES IN ONE PROBLEM

So far, we have seen how to use a molecular formula to calculate molar mass, and then how to use this molar mass as a conversion factor (between moles and mass). The previous problems focused on those calculations. But there are some problems that you might see that get a bit more difficult, even though they just require the same exact calculations. To illustrate the point, consider the following example:

How many grams of silver (Ag) could be obtained if we removed all of the silver from 47.2 g of silver nitrate ($AgNO_3$)?

When you first read this problem, it seems like a problem that requires the conversion of mass into mass (grams of one substance into grams of another substance). So on the surface, you might be tempted to say that you have not yet learned how to do this. But in fact, you have learned all the skills you need. In order to "see" how the problem is not so hard, you will have to train your eye to read problems in a certain way. Let's take a closer look.

We have said that if you are given the molecular formula, then you can always convert mass into moles, and moles into mass, by first calculating the molar mass and then using it as a conversion factor. Now that you have these skills, you must keep these skills in mind when reading a problem. When a problem says: "3.07 g of $MgCl_2$ . . .", you must immediately think to yourself like this: they gave me the molecular formula and the mass, so I can easily calculate how many moles I have. That is what you must think to yourself while you are reading any problem. Even if the problem isn't about that at all, you should still be thinking it. When it comes to solving problems, most of the battle is to realize what information you are given and how you can manipulate that information. Similarly, if you read a problem that says: "2.5 moles of NH_3 . . .", you should immediately think to yourself: they gave me the molecular formula and the moles, so I can easily calculate how many grams of NH_3 I have.

You have to train yourself to think in this way if you want to be able to solve the more difficult problems. And when you do train yourself, you will find that the more difficult problems are really not so difficult in the first place. Let's now revisit our "difficult" problem, and let's apply this new type of thinking strategy to try and solve this problem:

How many grams of silver (Ag) could be obtained if we removed all of the silver from 47.2 g of silver nitrate ($AgNO_3$)?

Let's analyze each half of the problem separately.

1. "How many grams of silver (Ag) could be obtained . . ."

As you read this, you should think to yourself like this: Whenever I have the molecular formula, I can easily interconvert grams and moles. (In this case, the molecular formula is provided—we are just talking about individual silver atoms, so the molecular formula is just Ag). You can look at the periodic table and see that the molar mass of silver is 107.9 g/mol. Therefore, *if* you can find out how many moles of silver are obtained, then you could easily convert your answer into grams.

Now let's look at the second part of the problem:

2. ". . . if we removed all of the silver from 47.2 g of silver nitrate ($AgNO_3$)?"

As you read this, you should think to yourself like this: Whenever I have the molecular formula, I can easily interconvert grams and moles. Since I know how many grams I started with, I can easily calculate how many moles I started with.

Now we can think about this problem in a new light. With the information that is given to us in the problem, we can easily calculate how many moles of silver nitrate we started with. And we have said that if we can calculate the number of moles of silver we produce, then we can convert that answer into grams of silver. So, this whole problem boils down to one simple question: how many moles of Ag are produced from one mole of $AgNO_3$?

If you have trouble seeing the logic we just used, try to stare at this diagram, which represents the logic visually:

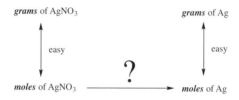

By using the concept of molar mass as a conversion factor, we can rephrase our question that makes the problem sound a lot simpler: how many moles of Ag are produced for every mole of $AgNO_3$? If we look at the molecular formulas, we can easily see that there is a 1:1 ratio here. Ag appears only once in $AgNO_3$, so 1 mole of $AgNO_3$ will produce 1 mole of Ag. Now we can solve our problem by using the molecular formula of $AgNO_3$ to calculate the molar mass of $AgNO_3$. Once we know the molar mass, we can use it to convert the *mass* of silver nitrate (47.2 g) into *moles* of silver nitrate. If we know how many moles of $AgNO_3$ we have, then we also know how many moles of Ag can be obtained (one mole of $AgNO_3$ will produce one mole of Ag atoms). Finally, if we know how many moles of silver we will form, then we can use the molar mass of silver to convert moles of silver into grams of silver.

Now think about the plan we just outlined for solving this problem. We convert grams into moles (for $AgNO_3$), and then we convert moles back into grams (for Ag). Essentially, we will be using molar mass as a conversion factor *twice*. Now let's use our plan to solve the problem.

EXERCISE 2.33.   How many grams of silver (Ag) could be obtained if we removed all of the silver from 47.2 g of silver nitrate ($AgNO_3$)?

Answer: Start by calculating the molar mass of $AgNO_3$. Use the periodic table to do this (looking at the numbers for Ag, N, and O). You will see that the molar mass of $AgNO_3$ = (107.9 g/mol) + (14.01 g/mol) + 3(16.00 g/mol) = 169.9 g/mol.

Therefore, 1 mole of $AgNO_3$ = 169.9 g of $AgNO_3$. Use this as one of your conversion factors in calculating your answer. In total, you need three conversion factors (look at the previous diagram). When you string these three conversion factors together, you get:

$$47.2 \text{ g } AgNO_3 \times \frac{1 \text{ mol } AgNO_3}{169.9 \text{ g } AgNO_3} \times \frac{1 \text{ mol Ag}}{1 \text{ mol } AgNO_3} \times \frac{107.9 \text{ g Ag}}{1 \text{ mol Ag}}$$

If you stare at the solution above, you will see all three conversion factors. The first one converts *grams* of silver nitrate into *moles* of silver nitrate. The second conversion factor is used to convert moles of *silver nitrate* into moles of *silver*, and the third conversion factor is used to convert *moles* of silver into *grams* of silver. When you calculate the above solution, you get: **30.0 g of Ag**

Now let's get some practice using the exact same procedure that we used above. You will need to use three conversion factors to get the answer. Try to set up the solution to this problem without looking back at the previous problem. If you get stuck, then look back to the previous problem and model your answer after that problem. Then, move on to the next problem, trying not to look back to formulate your solution. Keep going through the following problems until you can set up the solution without looking back.

PROBLEM 2.34. How many grams of silver could be obtained if we removed all of the silver from 12.07 g of silver chloride (AgCl)?

Molar mass of AgCl = _____

Solution = _____

PROBLEM 2.35. How many grams of calcium could be obtained if we removed all of the calcium from 2.20 g of calcium carbonate ($CaCO_3$)?

Molar mass of $CaCO_3$ = _____

Solution = _____

PROBLEM 2.36. How many grams of copper could be obtained if we removed all of the copper from 3.7 g of copper iodide (CuI)?

Answer: _____

PROBLEM 2.37. How many grams of silver could be obtained if we removed all of the silver from 6.75 g of silver oxide (Ag_2O)? (*Hint:* 1 mole of Ag_2O will yield 2 moles of silver atoms.)

Answer: _____

We have now observed that even apparently complex problems can be greatly simplified if you can train your eye to see when you can use conversion factors. If you want to do well on your exams, it is important that you train yourself to think in this way. And we are not just talking about molar mass, but we are talking about ANY conversion factor that you learned. So far, we have seen two special conversion factors that can be used to convert one type of unit into another: first we saw density in Chapter 1, and now we have seen molar mass. To make sure that we train our eyes to read problems in this way, let's just work through the following simple "fill-in-the-blank" problems:

PROBLEM 2.38. You are working on a problem that starts with the following words: "2.3 grams of silver sulfate (Ag_2SO_4) . . .". As you read this problem, you should think to yourself the following: "I have been given the molecular formula, which I can use to calculate the _____ With that, I can convert grams of silver sulfate into _____ of silver sulfate.

PROBLEM 2.39. You are working on a problem that starts with the following words: "9.7 moles of magnesium oxalate (MgC_2O_4) . . .". As you read this problem you should think to yourself the following: "I have been given the molecular formula, which I can use to calculate the _____ With that, I can convert moles of magnesium oxalate into _____ of magnesium oxalate.

2.7 COMBINING CONVERSION FACTORS

As you progress through this course, you will see problems where you need to use multiple conversion factors. To illustrate what we mean by this, consider the two special conversion factors that we have seen so far: *density* and *molar mass*. Let's quickly review what we have seen:

Density	allows us to interconvert	**mass** *and volume*
Molar mass	allows us to interconvert	**mass** *and moles*

Notice that mass can be interchanged in each of these conversions. So, if we put them together, we can actually convert moles into volume, or volume into moles, like this:

Let's see an example:

EXERCISE 2.40. At 25° C, the density of water is 1.00 g/cm^3. At this temperature, what volume will be occupied by 3.89 moles of water (H_2O)?

Answer: This problem is asking you to convert *moles* of water (3.89 moles) into a measurement of *volume*. To do this, you will need to do *two* conversions (as shown in the previous diagram). You are given the molecular formula of water, so you can easily calculate the molar mass (using the values on the periodic table, you get 18.02 g/mol), which you can use to convert moles into grams. Then you are given the density, which you can use to convert from grams into volume (g/cm³). So, your solution will have two conversions:

$$3.89 \text{ mol} \times \frac{18.02 \text{ g}}{1 \text{ mol}} \times \frac{1 \text{ cm}^3}{1.00 \text{ g}}$$

For each of these conversions, you just need to consider which fraction to use (which units go into the numerator and which go in the denominator) in order that your units should all cross off in the end (you practiced this Chapter 1). After doing the calculation above, your answer should be **70.1 cm³**.

PROBLEM 2.41. At 20° C, the density of ethanol (C_2H_5OH) is 0.789 g/cm³. At this temperature, what volume will be occupied by 1.50 moles of ethanol?

Answer: _____

PROBLEM 2.42. At 20° C, the density of gold (Au) is 19.3 g/cm³. At this temperature, what volume will be occupied by a bar of gold made of 9.5 moles of gold atoms?

Answer: _____

In the previous problems, we saw how to use two special conversion factors (density and molar mass) to convert from *moles* into *volume*. In these problems, we used the molar mass to convert *moles* into *grams*, and then we use density to convert *grams* into *volume*:

Similarly, we can do the exact same process in the other direction. We can convert *volume* into *moles* using the same two conversion factors. We would start by converting *volume* into *grams* (using density), and then by converting *grams* into *moles* (using molar mass):

Let's see an example:

EXERCISE 2.43. The density of ethylene glycol ($C_2H_6O_2$) at 25° C is 1.09 g/cm³. If you have a 100.0 cm³ container filled with ethylene glycol, how many moles of ethylene glycol are in the container?

Answer: In this problem, you must convert volume into moles. So two conversion factors are needed: density and molar mass. You were given the density, and you will have to calculate the molar mass from the molecular formula. The molar mass of ethylene is:

$$2(12.01 \text{ g/mol}) + 6(1.008 \text{ g/mol}) + 2(16.00 \text{ g/mol}) = 62.07 \text{ g/mol}$$

Now you have both conversion factors that you need. In order to convert volume into moles, you will need to first convert volume into grams, and then grams into moles. So set up the following solution:

$$100.0 \text{ cm}^3 \times \frac{1.09 \text{ g C}_2\text{H}_6\text{O}_2}{1 \text{ cm}^3} \times \frac{1 \text{ mol C}_2\text{H}_6\text{O}_2}{62.07 \text{ g C}_2\text{H}_6\text{O}_2}$$

And you get: **1.76 moles of C$_2$H$_6$O$_2$.**

PROBLEM 2.44. The density of ethanol (C$_2$H$_5$OH) at 25° C is 0.79 g/cm^3. If you have a 10.5 cm^3 container filled with ethanol, how many moles of ethanol are in the container?

Answer: _____

PROBLEM 2.45. The density of gold (Au) at 25° C is 19.32 g/cm^3. If you have a bar of gold that occupies a volume of 3.4 cm^3, how many moles of gold do you have?

Answer: _____

PROBLEM 2.46. The density of iron (Fe) at 25° C is 7.9 g/cm^3. If you have a bar of iron that occupies a volume of 5.2 cm^3, how many moles of iron do you have?

Answer: _____

2.8 MASS PERCENT AND ELEMENTAL ANALYSIS

In the beginning of this chapter, we saw the difference between an empirical formula and a molecular formula. At the time, we saw that a molecular formula is more useful because it tells us the exact number of each atom in the molecule. An empirical formula only tells us the ratio of atoms in the molecule. You might wonder why we use empirical formulas at all if they are less useful? It turns out that we can easily determine the empirical formula of an unknown compound by running one simple experiment called *elemental analysis*. In this section, we will see how to use the data from an elemental analysis to determine an empirical formula.

An elemental analysis tells us the amount of grams of each type of atom present in the molecule. We will soon see how to use this information, but first we need to focus on a term called *mass percent* (or mass %). To understand what mass %

is, let's go back to our Lego analogy. Imagine that we went to the Lego warehouse and we bought one barrel containing the following structures:

On our way out of the warehouse, we weigh our barrel and we find that the barrel weighs 100 lb. Suppose we wanted to know how much of this weight comes from the black pieces? It might be tempting to say this: there are three Lego pieces in every structure, and only one of them is black, so black pieces should be one third of the total weight (or 33 lb). The problem with that argument is that black pieces and white pieces have different weights. And we have to take that into account.

So we do it like this. We take our barrel (which has 6,022,042 structures in it), and we dump the contents onto the floor. Then we break apart all of the structures into individual Lego pieces, and we get over 12 million white pieces and over 6 million black pieces. We then sort the pieces into two piles; one pile of white pieces and one pile of black pieces. When we weigh all of the white pieces together, we get 20 lb. When we weigh all of the black pieces together, we get 80 lb. Now we have our answer. The *mass* % of the black pieces is therefore:

$$\frac{\text{mass of black pieces}}{\text{total mass}} \times 100\% = \frac{80 \text{ lb}}{100 \text{ lb}} \times 100\% = 80\%$$

We can say that the black pieces contributed 80% of the weight of the original barrel, and similarly, the white pieces contributed 20% of the weight of the original barrel. That's what we mean by mass %.

Now let's do the same thing with water molecules instead of Lego pieces:

$$H \diagdown \underset{\cdot\overset{\cdot}{O}\cdot}{} \diagup H$$

If we take 1 mole of water, and we measure its mass, we see that it is 18.02 grams. If we want to know how much of this mass comes from oxygen atoms, we need to look at the mass of 1 mole of oxygen atoms. One mole of oxygen atoms is 16.00 grams. So, the mass % of oxygen atoms in 1 mole of H_2O is:

$$\frac{\text{mass of oxygen atoms}}{\text{total mass}} \times 100\% = \frac{16.00 \text{ g}}{18.02 \text{ g}} \times 100\% = 88.79\%$$

If we did the same kind of calculation for hydrogen, we would find that the mass % of hydrogen in 1 mole of H_2O is 11.21%. This should make sense because 88.79% + 11.21% = 100%, which is what we would expect. Sometimes rounding the individual mass percentages can cause the sum total to be 99.99%. Don't worry about that—it is just an artifact of rounding the numbers, which is what we have to do when we report numbers with the appropriate significant figures.

The concept of mass % is a straightforward concept, but it only seems that way when you have the molecular formula in front of you. Suppose you have an unknown compound and you are given the percentage composition (that is just a list of the mass % of each element in the compound). Will you be able to reconstruct an empirical formula? What about a molecular formula? We will focus on these topics in just a few moments. For now, let's just do some simple exercises to make sure that you are comfortable with the concept of mass %.

EXERCISE 2.47.   Consider the compound called hydrazine (N_2H_4). What is the mass % of nitrogen in this compound?

Answer: First you must determine the molar mass, which can easily be derived from the molecular formula: $2(14.01 \text{ g/mol}) + 4(1.008 \text{ g/mol}) = 32.05 \text{ g/mol}$. So, 1 mole of hydrazine has a mass of 32.05 g (or we can say that 1 *molecule* of hydrazine has a mass of 32.05 u).

Next, we look at how much of this mass comes from nitrogen:

$$\frac{\text{mass of nitrogen atoms}}{\text{total mass}} \times 100\% = \frac{2(14.01) \text{ g}}{32.05 \text{ g}} \times 100\% = 87.43\%$$

Notice that we could have done this calculation in atomic mass units instead of grams (essentially looking at 1 molecule instead of 1 mole), and we would get the exact same answer.

PROBLEM 2.48.   Calculate the mass % of calcium in calcium chloride ($CaCl_2$):

PROBLEM 2.49.   Calculate the mass % of chlorine in calcium chloride ($CaCl_2$). Make sure that your answer to this problem plus your answer to the previous problem = 100%.

PROBLEM 2.50. Calculate the mass % of zinc in zinc sulfide (ZnS):

PROBLEM 2.51. Calculate the mass % of carbon in butanoic acid ($C_4H_8O_2$):

PROBLEM 2.52. Calculate the mass % of oxygen in oxalic acid ($C_2H_2O_4$):

    In each of these problems, we knew the molecular formula, and we were able to use that information to calculate mass percentages. More often, we deal with *unknown* compounds, where we don't know what the molecular formula is. This is where elemental analysis becomes important. An elemental analysis is an experiment that tells us how many grams of each element were present in the original compound. With this information, it is actually easier to calculate the percentage composition. For example, suppose we have an unknown sample with a mass of 33.55 g. An elemental analysis produces 13.42 g of carbon, so we can calculate the mass % of carbon like this:

$$\frac{13.42 \text{ g}}{33.55 \text{ g}} \times 100\% = 40.00\%$$

This is the easiest way to calculate mass %. We don't have to bother with the molecular formula or with calculating molar mass, like we did earlier. We just use the information from an elemental analysis and we're done. Since it's so easy, you probably won't see too many problems like this, but there is a common type of problem that you will definitely see. When given the information from an elemental analysis, you will see problems where you must determine an empirical formula.

The procedure for doing this is not difficult, because it just involves a procedure that we have already seen and mastered: converting grams into moles. We have already seen how to convert grams into moles using molar mass as a special conversion factor. Now, we don't know the molar mass of the unknown compound, because we don't know its molecular formula. But, we do know the molar mass of individual elements. All we have to do is look at the periodic table for that. For example, we have already seen that a single carbon atom will have a mass of 12.01 u, and a mole of carbon atoms will have a mass of 12.01 grams. So, the molar mass of carbon atoms is just 12.01 g/mol. We can use that molar mass to convert *grams* of carbon into *moles* of carbon. And that is the whole technique behind solving these problems. Let's see an example of how we do it.

EXERCISE 2.53. A sample was found to contain 13.42 g of carbon, 2.25 g of hydrogen, and 17.88 g of oxygen. What is the empirical formula for this compound?

Answer: Whenever you are given the information from an elemental analysis (amounts of elements listed in grams), you will have to convert each of these amounts into moles. To do this, you will need the molar mass for carbon atoms, for hydrogen atoms, and for oxygen atoms. You get each of these values directly from the periodic table. The molar mass of carbon is 12.01 g/mol, the molar mass of hydrogen is 1.008 g/mol, and the molar mass of oxygen is 16.00 g/mol. Now, you can convert all of the data you were given into moles, using the factor-label method:

$$13.42 \text{ g carbon} \times \frac{1 \text{ mol carbon}}{12.01 \text{ g carbon}} = 1.117 \text{ mol carbon}$$

$$2.25 \text{ g hydrogen} \times \frac{1 \text{ mol hydrogen}}{1.008 \text{ g hydrogen}} = 2.23 \text{ mol hydrogen}$$

$$17.88 \text{ g oxygen} \times \frac{1 \text{ mol oxygen}}{16.00 \text{ g oxygen}} = 1.118 \text{ mol oxygen}$$

Now you know how many moles of each element were produced in the elemental analysis. You can summarize the ratio like this:

$$C_{1.117}H_{2.23}O_{1.118}$$

But you cannot report your answer like this, because as you'll recall, an empirical formula uses the smallest whole digits possible as subscripts. The subscripts above are not whole numbers. You might notice right away that the ratio above is just

CH_2O, but sometimes you might not notice the ratio so readily. So, let's see the procedure you can use whenever you get stuck.

The trick goes like this: just choose the smallest number, and divide all other numbers by that number. In this case, the smallest number is 1.117. If you divide all of these numbers by 1.117, you get:

$$\frac{1.117}{1.117} = 1.000 \qquad \frac{2.23}{1.117} = 2.00 \qquad \frac{1.118}{1.117} = 1.001$$

So, with the ratio of 1:2:1, that is how you get:

$$CH_2O$$

Notice that, in the example above, you were only able to calculate the *empirical* formula. We still don't know what the molecular formula is. If the empirical formula is CH_2O, then the molecular formula could be CH_2O or $C_2H_4O_2$ or $C_3H_6O_3$. . . . In order to determine which molecular formula you have, you will need one more piece of information. Before we do this, though, let's get practice doing problems like the example above.

Remember that the technique is to convert all the data from grams into moles. To do this you will use the molar mass of each element, which you can get from the periodic table. Once you convert everything into moles, divide all numbers by the smallest number, and you will have the ratio you need for your empirical formula.

PROBLEM 2.54.   A sample was found to contain: 50.44 g of carbon, 12.70 g of hydrogen, and 33.60 g of oxygen. What is the empirical formula for this compound?

Answer: _____

PROBLEM 2.55.   A sample was found to contain: 43.29 g of sulfur and 64.8 g of oxygen. What is the empirical formula for this compound?

Answer: _____

PROBLEM 2.56.   A sample was found to contain: 8.982 g of phosphorus and 0.8770 g of hydrogen. What is the empirical formula for this compound?

Answer: _____

Sometimes, you will see a problem just like the problems above, BUT there will be one minor difference. Instead of giving you the data from elemental analysis *in grams* (X grams of carbon, Y grams of oxygen, etc.), the problem will give you a *percentage composition* (mass % of carbon is X%, mass % of oxygen is Y%, etc.). When you see a problem like this, there is a simple trick to put this problem into the form of all the problems above. All you do is pretend that you started with 100 g of sample. In this case, 62.5% carbon is the same as 62.5 grams of carbon.

By pretending to start with a sample size of 100 grams, you are essentially converting percentage composition into mass composition. Let's see an example:

EXERCISE 2.57.   A sample has the following percentage composition: 40.00% carbon, 6.71% hydrogen, and 53.29% oxygen. What is the empirical formula for this compound?

Answer: In this problem, you are not given the actual grams of carbon, hydrogen, and oxygen. So, it seems like a different type of problem then what we are used to. Here is the trick that we just mentioned. The percentages above are true for any amount of substance, whether it is 10 grams, or 35 grams, or 100 grams of sample. If we imagine a sample size of 100 grams, then the problem becomes much simpler. 40.00% carbon becomes 40.00 grams of carbon. 6.71% hydrogen becomes 6.71 grams of hydrogen. And 53.29% oxygen becomes 53.29 grams of oxygen. And now the problem is in exactly the same format as all of the problems we just did.

Now that each of the elements is measured in *grams*, you will have to convert each of these amounts into *moles*. To do this, you will need the molar mass for carbon atoms, hydrogen atoms, and oxygen atoms. You get each of these values directly from the periodic table. The molar mass of carbon is 12.01 g/mol, the molar mass of hydrogen is 1.008 g/mol, and the molar mass of oxygen is 16.00 g/mol. You can now convert all of the data were given into moles, using the factor-label method:

$$40.00 \text{ g carbon} \times \frac{1 \text{ mol carbon}}{12.01 \text{ g carbon}} = 3.331 \text{ mol carbon}$$

$$6.71 \text{ g hydrogen} \times \frac{1 \text{ mol hydrogen}}{1.008 \text{ g hydrogen}} = 6.66 \text{ mol hydrogen}$$

$$53.29 \text{ g oxygen} \times \frac{1 \text{ mol oxygen}}{16.00 \text{ g oxygen}} = 3.331 \text{ mol oxygen}$$

Now you know how many moles of each element were produced in the elemental analysis. You can summarize the ratio like this:

$$C_{3.331}H_{6.66}O_{3.331}$$

But you cannot report your answer like this because, as you'll remember, an empirical formula uses the smallest whole digits possible as subscripts. The ratio is 1:2:1, so your empirical formula is:

$$CH_2O$$

To make sure you understand the trick, try to do a problem that gives you elemental analysis data in the format of percentage composition.

PROBLEM 2.58.   A sample has the following percentage composition: 50.05% sulfur and 49.95% oxygen. What is the empirical formula for this compound?

Answer: _____

In all of the problems we have just seen, we have been able to reconstruct an empirical formula from experimental data (elemental analysis) of an unknown compound. But what if we wanted to reconstruct a molecular formula? We have already seen that a molecular formula is much more useful. It turns out that we only need one more piece of information to determine the molecular formula. We need to be given the molar mass. If a problem gives us elemental analysis data AND a molar mass of the unknown compound, then we can determine the molecular formula. To see how this works, consider one of the problems that we recently did.

In exercise 2.53 (just a few pages ago), we were given the data from an elemental analysis, and we used those data to determine that the empirical formula was CH_2O. And that was where we left off. At the time, there was no way for us to tell if we were talking about CH_2O or $C_2H_4O_2$ or $C_3H_6O_3$ . . . etc. All of these formulas have the same ratio of carbon to hydrogen. Since we don't know which one it is, we write the empirical formula as the simplest whole numbers that show this ratio. In this case, that is CH_2O. But if we had been given the molar mass, then we would have been able to determine which one is the correct molecular formula. Let's see how this works:

EXERCISE 2.59.   In Exercise 2.53, you determined that the empirical formula was CH_2O. The molar mass of this compound is 60.05 g/mol. What is the molecular formula?

Answer: If you calculate the molar mass of the compound CH_2O, you find that it is 30.03 g/mol. The compound you want is 60.05 g/mol, so you have to multiply all of your subscripts by a factor of two. When you do this, you get:

$$C_2H_4O_2$$

That's all there is to it.

Now let's combine it all together in the following three final problems for this chapter.

PROBLEM 2.60.   A sample was found to contain: 2.270 g of carbon and 0.19105 g of hydrogen. The molar mass of the compound is 78.11 g/mol. What is the molecular formula for this compound?

PROBLEM 2.61. A sample was found to contain: 1.084 g of hydrogen, 15.06 g of nitrogen, and 34.40 g of oxygen. The molar mass of the compound is 47.01 g/mol. What is the molecular formula for this compound?

PROBLEM 2.62. A sample was found to contain: 1.062 g of hydrogen, 133.757 g of iodine, and 50.589 g of oxygen. The molar mass of the compound is 175.91 g/mol. What is the molecular formula for this compound?

CHAPTER 3

BALANCING REACTIONS AND STOICHIOMETRY

In this chapter, we will deal with a topic called *stoichiometry*, which is the study of *mass relationships* in chemical reactions. To simplify what we mean by mass relationships, let's consider a reaction where A reacts with B to form C:

$$A + B \longrightarrow C$$

The following questions belong to the realm of stoichiometry: How many moles of A are necessary to completely react with 2 *moles* of B? What mass of A is necessary to completely react with 20 *grams* of B? what mass of C will be formed if you start with 10 grams of A and 12 grams of B? All of these questions, and others like them, describe the relationships between amounts of chemicals participating in the chemical reaction (whether we measure in moles or grams).

We will begin the chapter with the skills necessary for balancing reactions. These skills are critical, because stoichiometry problems usually *build upon* your ability to balance a reaction. If you can't balance a reaction, then you won't have a starting point to do any other kind of stoichiometry problem.

Once we have mastered the skills involved in balancing reactions, we will then turn our attention to the most common types of stoichiometry calculations (like the questions above).

3.1 CONSERVATION OF MASS: COUNTING ATOMS ON BOTH SIDES OF A REACTION

The guiding principle behind balancing reactions is *conservation of mass*. This means that atoms can neither be created nor destroyed during a reaction. To see this more clearly, let's go back to our Lego analogy from Chapter 2.

In our analogy, we can combine individual pieces to form structures:

Consider the instructions shown above for building a specific Lego structure. We can see that we need white pieces and black pieces to build the structure. But the instructions above are "unbalanced". There is only one white piece on the left side of the instructions, and there are three white pieces on the right side. We could *balance* the instructions like this:

Now imagine that you were being paid to assemble two of these structures. You can deduce from the "balanced" instructions above that you will need six white pieces and two black pieces:

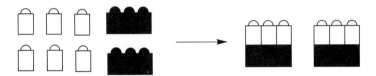

If someone gave you six white pieces but only **one** black piece, then you would not have enough to build two structures. Also, if someone gave you six white pieces and **three** black pieces, then you would have *too many* black pieces. You would have one black piece left over in the end. In order to build *exactly* two structures and have *no leftover* pieces, then you need a very specific number of starting pieces.

The same concept is true if we are building structures in large amounts. For example, if you had to build two *barrels* of structures, you would need six barrels of white pieces and two barrels of black pieces. You couldn't do your job if you only had five barrels of white pieces. And if you had seven barrels of white pieces, then you would have one barrel of white pieces left over when you finished your job (and you would have a problem of what to do with the leftover barrel).

And that brings us to the basic idea behind conservation of mass. In our analogy, you cannot create Lego pieces out of nothing, and you cannot make extra pieces disappear when you are done with the job. If you want to make sure that you have enough pieces to create your structures, and if you want to make sure that you don't have any leftover pieces when you are done, then you will need to have the correct number of starting pieces.

So far, not so difficult. Now, let's complicate things a bit. You call the Lego warehouse, and they tell you that they don't sell barrels of individual white pieces. They only sell them as white *structures* (where one structure = two white pieces stuck together):

Now, remember how barrels are sold. Every barrel is sold with the same number of *structures* (just over 6 million). So, if we buy one barrel of white *structures* and then take apart the structures into individual white pieces, then we will have over

12 million white pieces. So, now we don't need three white barrels anymore. In fact, if we buy three barrels of white *structures*, we will have twice as many white pieces as we need. So, we only need $1^1/_2$ barrels of white structures. Since $1^1/_2$ is the same as $^3/_2$, we can write it like this:

This equation is **balanced**, because we have the same number of white pieces on either side, and we have the same number of black pieces on either side.

Now we add one more complication. You tell the Lego warehouse that you want one barrel of black pieces and $1^1/_2$ barrels of white structures, and they tell you that they don't sell half barrels. They only sell complete barrels. You tell your boss what is going on, and you are told that you cannot have any leftover pieces when you are done. If you buy two barrels of white structures, you will have more than you need, and you will have leftover pieces. So, your boss tells you to buy three barrels of white structures and two barrels of black pieces. This is essentially just the equation above times 2:

This is also a balanced equation, but this time, the recipe does not involve any half barrels.

Now let's do the exact same example again, but this time we will use moles of atoms and molecules, instead of barrels of Lego pieces and structures. Let's say that the white pieces are chlorine atoms and the black pieces are aluminum atoms. So, we want to combine these two types of atoms to make a mole of $AlCl_3$ molecules:

$$Cl + Al \longrightarrow AlCl_3$$

The problem is that this reaction is not balanced. To balance it, we will need 3 moles of chlorine atoms:

$$3Cl + Al \longrightarrow AlCl_3$$

Now the reaction is balanced, but there is one problem. When you try to find chlorine atoms in nature, you find that chlorine exists (under standard conditions) as diatomic molecules (Cl-Cl). This is the same idea as when we found out that white pieces are only sold as structures, rather than individual pieces. So, *1* mole of chlorine *molecules* is essentially equal to *2* moles of chlorine *atoms*. We don't need 3 moles of chlorine anymore—that would be too much. We only need three halves of a mole of chlorine:

$$^{3}/_{2} \ Cl_2 + Al \longrightarrow AlCl_3$$

Now the reaction is ***balanced***. But, we have a fraction as a coefficient ($^{3}/_{2}$). In general, we try to avoid that, and we use whole numbers as coefficients. The truth is that nature is different than the Lego warehouse, because nature *will* provide us with half of a mole, or three halves of a mole. So the reaction above is a perfectly valid, balanced reaction. But we prefer to multiply everything by some number that gives us the smallest possible ***whole number*** coefficients. In this case, we multiply by 2:

$$3 \ Cl_2 + 2 \ Al \longrightarrow 2 \ AlCl_3$$

The reason we prefer whole numbers is simple. The reaction above can be read in one of two ways:

- 3 *moles* of Cl_2 reacts with 2 *moles* of Al to give 2 *moles* of $AlCl_3$, or
- 3 *molecules* of Cl_2 reacts with 2 *molecules* of Al to give 2 *molecules* of $AlCl_3$.

So, we can refer to ***molecules*** reacting, or entire ***moles*** reacting. It might make sense to talk about half of a *mole*, but it does not make sense to talk about half of a *molecule*. To preserve the dual meaning of the reaction (molecule or moles), we try to use whole numbers. When you learn about thermodynamics (later in this course), you will see that it is often desirable to write down a reaction showing the formation of exactly *1* mole of product. In situations like this, it is common to have a reaction where some of the coefficients are fractions. For example, it is common to see examples of $^{1}/_{2} \ O_2$ in balanced reactions. Therefore, it is not the end of the world if you use coefficients that are fractions. The most important thing is that your reaction is balanced (same number of atoms on either side). But let's get in the habit of always using whole numbers as coefficients.

Now we are ready to start working on some problems. To slowly develop the skill you need for balancing reactions, we will start off with problems similar to the reaction we saw above between Al and Cl_2 to form $AlCl_3$. In that reaction, we had one element reacting with another element to form a compound that contained both elements. So, all you have to do is make sure you have the right coefficients in front of the elements on the left side of the reaction, and then you multiply by some coefficient that gets rid of any fractions.

As you are doing these problems, there are a few rules to keep in mind:

1. Never change a subscript on a compound. For example, if you are dealing with the compound NO_2, then you cannot change the subscripts to N_2O_2 or N_2O. Those are different compounds. You are only allowed to change *coefficients* in front of the compound. Your final answer should show the term NO_2.

2. Coefficients apply to everything after them. So, $3N_2O_4$ means that we have a total of 6 nitrogen atoms and 12 oxygen atoms. Sometimes we use parentheses in molecular formulas, like this: $CO(NH_2)_2$. The subscript after the parentheses tells us that we have two NH_2 groups, which means that we have 2 nitrogen atoms and 4 hydrogen atoms. When we add a coefficient, we multiply the coefficient by these numbers. So, $3CO(NH_2)_2$ tells us that we have 3 carbon atoms, 3 oxygen atoms, 6 nitrogen atoms, and 12 hydrogen atoms.

3. Never add or delete reagents. When you are given an unbalanced reaction that you need to balance, you just need to change the coefficients. That's it. Don't add in any new reagents and don't get rid of any reagents to make the reaction balance. It is very tempting at times to add O_2 or H_2 to the reaction to make it simpler to balance, but you are changing the reaction if you do that.

Now let's see an example:

EXERCISE 3.1. Balance the following reaction:

$$Ca + C + O_2 \longrightarrow CaCO_3$$

Answer: Once again, this is the simplest type of problem there is. The left side of the reaction has only individual elements, and the right side has only one compound containing all of these elements. So, we start by placing coefficients on the left side to balance all of the elements.

We see that we have one calcium atom on either side, so the calcium atoms are already balanced. We also see that we have one carbon atom on either side, so the carbon atoms are already balanced. But, we have two oxygen atoms on the left side, and three on the right side. To get a total of three oxygen atoms on the left side, we would need $^3/_2 O_2$:

$$Ca + C + {}^3/_2 O_2 \longrightarrow CaCO_3$$

Now we have a balanced reaction. Our final step is to multiply by the smallest number that will turn all of our coefficients into whole numbers. In this case we multiply by 2 (if we multiplied by 4, or by 8, we would also get whole numbers, but we try to get the *smallest* whole numbers possible), and we get:

$$2\ Ca + 2\ C + 3\ O_2 \longrightarrow 2\ CaCO_3$$

PROBLEM 3.2. Balance the following reaction by placing the appropriate coefficient in front of each substance in the reaction. Start by placing coefficients on the left side to balance all of the elements (even if some of the coefficients are fractions). Then multiply by a number that will make all of the coefficients whole numbers.

$$\_\_Na + \_\_H_2 + \_\_C + \_\_O_2 \longrightarrow \_\_NaHCO_3$$

PROBLEM 3.3. Balance the following reaction by placing the appropriate coefficient in front of each substance in the reaction:

$$\_\_Al + \_\_O_2 \longrightarrow \_\_Al_2O_3$$

PROBLEM 3.4. Balance the following reaction by placing the appropriate coefficient in front of each substance in the reaction:

$$\_\_H_2 + \_\_N_2 + \_\_O_2 \longrightarrow \_\_HNO_3$$

PROBLEM 3.5. Balance the following reaction:

$$F_2 + H_2 \longrightarrow HF$$

PROBLEM 3.6. Balance the following reaction:

$$Cs + Cl_2 \longrightarrow CsCl$$

PROBLEM 3.7.   Balance the following reaction:

$$Cu + S_8 \longrightarrow Cu_2S$$

3.2 RULES AND TOOLS FOR BALANCING REACTIONS

So far, we have seen how to balance simple reactions where we have *elements* on the left side of the reaction, and compounds (molecules consisting of more than one kind of atom) on the right side of the reaction. These problems are the simplest kinds of problems. More often, you will see compounds on *both* sides of the reaction. For example:

$$NH_4NO_3 \longrightarrow N_2 + O_2 + H_2O$$

Examples like this can be a bit more complex than the problems we have seen so far. Our method in the previous problems was to balance all elements on the left side of the reaction, and then we were done (in some cases, we just had to multiply by some number to get rid of fractions). But now, we will be dealing with cases where we need to play around with the coefficients on *both sides* of the reaction. We will have to act like a ping-pong ball, bouncing back and forth from one side of the reaction to the other. Sometimes, it can take a while to balance a reaction. So, to make things go faster, there are a couple of tools that will help you.

Tool #1.   Before getting started, always look at the reaction and locate any reagents that consist of only one type of atom. For example: O_2, or H_2, or Zn, or Al, or O_3, etc. If you inspect each of these examples, you will see that they all have one thing in common. They all possess only one kind of atom, whether there is a subscript (like in O_3) or not (like in Zn). Once you have identified these reagents, you should realize that those elements are the easiest to balance, because you can always put a coefficient in front of one of these reagents without affecting anything else. Therefore, we will balance these reagents LAST. Let's take a closer look at why we do this.

When you have a reagent that consists of many different elements (such as $NaHCO_3$ or C_3H_6O), then you really change a lot when you put a coefficient in front of that reagent. For example, if you place a 2 in front of $NaHCO_3$ on one side of a reaction, then you will have to go to the other side of the reaction and balance all of the elements (Na, H, C, and O). Then, when you try to make that change on the other side, you find that you need to go back like a ping-pong ball from one side of the reaction to the other. At times, this can send you into a frustrating sequence of back-and-forth balancing that will take a long time. However, when you place a 2 in front of Zn or H_2, you don't affect anything else. That's why we save these elements for last when we are balancing a reaction.

It is always a good idea to start balancing a reaction by simply writing a list of all the elements that need to be balanced. Then, you can identify any elements on the list that you will balance last. To illustrate this, let's work through the example we saw in the beginning of this section:

$$NH_4NO_3 \longrightarrow N_2 + O_2 + H_2O$$

If we prepare a list of all of the elements that we see in this reaction, we get:

N, H, and O.

Then, we look to see if any of these appear by themselves in the reaction above. And we find that both N and O appear by themselves on the right side of the reaction (N_2 and O_2). This tells us to focus on N and O last, which means we must start by balancing the hydrogen atoms. If we can do that, then we can come back and easily balance N and O.

To balance the hydrogen atoms, we focus on the reagents that contain H on both sides of the reaction. We have one reagent on the left side of the reaction (NH_4NO_3) that has four hydrogen atoms, and one reagent on the right (H_2O) that has two hydrogen atoms. So, we put a 2 in front of the reagent on the right side:

$$NH_4NO_3 \longrightarrow N_2 + O_2 + \mathbf{2}\ H_2O$$

Now the hydrogen atoms are balanced. We just need to balance the nitrogen and oxygen atoms, which we can do by placing the coefficients in front of N_2 and O_2 and that will balance the reaction. Let's go ahead and do it for this example, because this is an example where we will end up using fractions.

Let's focus on nitrogen first (it doesn't really matter whether we do N or O first). There are two nitrogen atoms on the left side of the reaction, so we need two on the right side. We already have two nitrogen atoms on the right side, so the nitrogen atoms are already balanced.

Now let's focus on balancing the oxygen atoms. There are three oxygen atoms on the left side of the reaction, so we need three on the right side. We already have two oxygen atoms in $2H_2O$, so we need one more oxygen atom. To get one more, we write $^1/_2$ as a coefficient in front of O_2, and we get:

$$NH_4NO_3 \longrightarrow N_2 + \mathbf{^1/_2}\ O_2 + \mathbf{2}\ H_2O$$

Fractions are what allow us to balance odd numbers of atoms. So, if we had decided that we needed to add three oxygen atoms, then we would have used the coefficient $^3/_2$ in front of O_2. If we had decided that we needed 13 oxygen atoms to balance the reaction, then we would have used the coefficient $^{13}/_2$ in front of O_2.

Now we have a balanced reaction. But one of our coefficients is a fraction, so we multiply all of the coefficients by 2, and we get:

$$2\ NH_4NO_3 \longrightarrow 2\ N_2 + O_2 + 4\ H_2O$$

EXERCISE 3.8.   Balance the following reaction:

$$N_2H_4 + O_2 \longrightarrow NO_2 + H_2O$$

Answer: Begin any balancing problem by making a list of all elements that need to be balanced. If you look on the left side of the reaction, you see that there are three types of elements in this reaction:

N, H, and O.

If you look at the right side, you will find exactly the same three elements (if not, then there is a major problem because no coefficients could possibly balance a reaction with different types of elements on either side).

It is helpful to take a few seconds and make this list of elements, because it forces you to consider what it takes to balance the reaction. In order for this reaction to be balanced, you need to make sure that you have the same number of N on both sides of the reaction, the same number of H on both sides, and the same number of O on both sides.

Next, identify if any of these three elements is present by itself in the reaction (as a reagent consisting of only one element). In this reaction, oxygen appears by itself on the left side:

$$N_2H_4 + \boxed{O_2} \longrightarrow NO_2 + H_2O$$

So, keep in mind that you will balance the oxygen atoms *last* when you try to balance this reaction. Therefore, you will start by balancing N and H.

On the left side of the reaction, you have two nitrogen atoms, and on the right side you only have one nitrogen atom. So, place a 2 in front of NO_2 on the right side, and this balances the nitrogen atoms:

$$N_2H_4 + O_2 \longrightarrow 2\ NO_2 + H_2O$$

Notice that placing this 2 in front of NO_2 will change the amount of oxygen atoms, but it doesn't matter because you have already decided that it will be easy to balance oxygen last.

Now, you need to balance the hydrogen atoms. You have four hydrogen atoms on the left side, and only two hydrogen atoms on the right side. To balance this, place a 2 in front of H_2O:

$$N_2H_4 + O_2 \longrightarrow 2\ NO_2 + 2\ H_2O$$

Once again, you have changed the amount of oxygen atoms by placing the 2 in front of H_2O. If you had tried to balance the oxygen after you balanced the nitrogen atoms, then you would just have to rebalance the oxygen atoms again now. So, once again, you can see the power of choosing the oxygen to be balanced last, which is now all that is left to do.

To balance the oxygen atoms, count the oxygen atoms on both sides. There are two oxygen atoms on the left side, and six on the right side. In order to balance the oxygen atoms, you will need to have six on the left side, so change the coefficient to 3:

$$N_2H_4 + 3\ O_2 \longrightarrow 2\ NO_2 + 2\ H_2O$$

There are no fractions as coefficients, so you are done.

Now let's get a bit of practice. For each of the following unbalanced reactions, identify the smallest whole number coefficients necessary to balance the reaction. Remember to start by making a list of the elements that need to be balanced. Then determine which element(s) should be balanced *last*:

PROBLEM 3.9.

$$Cr_2O_3 + Mg \longrightarrow Cr + MgO$$

PROBLEM 3.10.

$$CH_4 + H_2O \longrightarrow CO + H_2$$

PROBLEM 3.11.

$$H_2CO + O_2 \longrightarrow CO + H_2O$$

PROBLEM 3.12.

$$NO + O_2 \longrightarrow NO_2$$

PROBLEM 3.13.

$$NO_2 + F_2 \longrightarrow NO_2F$$

PROBLEM 3.14.

$$CH_3NH_2 + O_2 \longrightarrow CO_2 + H_2O + N_2$$

PROBLEM 3.15.

$$Fe + H_2O \longrightarrow Fe_3O_4 + H_2$$

Tool #2. In the previous tool, we saw which reagents to save for LAST when balancing a reaction. Now, we focus on which elements to START with. Always start with the elements that are present in the compound with the largest number of different types of elements. Let's see an example of how this works:

EXERCISE 3.16. Balance the following reaction:

$$COCl_2 + H_2O \longrightarrow CO_2 + HCl$$

Answer: Begin by making a list of the elements that you need to balance. There are many elements in this reaction:

C, O, Cl, and H

You will need to balance these four elements on both sides of the reaction.

Next, look to see if there are any elements that appear by themselves in the reaction; there aren't any that appear alone. All of the reagents in this reaction are compounds. So, Tool #1 doesn't really help. What you need now is a good place to start. And that is where Tool #2 can help.

You can usually get to your answer faster if you choose the compound that contains the largest number of different elements. In the example that would be $COCl_2$, which is comprised of three different elements: C, O, and Cl. So, start with those three elements. Next, you need to decide which of those three elements to start with. To do that, look at how many places each element appears in the reaction. This reaction has four compounds: two on the left and two on the right. Carbon and chlorine each appear in two different compounds. Oxygen appears in three compounds, so it will probably be the toughest. So, start with C and Cl.

The carbon atoms are already balanced, and you can balance the chlorine atoms by placing a 2 in front of HCl:

$$COCl_2 + H_2O \longrightarrow CO_2 + \mathbf{2}\ HCl$$

Now you have balanced C and Cl. But you still have to balance O and H. When you look at the reaction, you will see that you have two hydrogen atoms on each side of the reaction, and you have two oxygen atoms on each side of the reaction. So, you are done.

The problem was fairly quick to solve, because you chose the right starting place: the compound with the largest number of different elements.

In each of the following problems you will need to go back and forth like a ping-pong ball a few times. None of these problems will require you to go back and forth *too* many times (we will soon see what to do in those cases). For now, let's focus on problems where we have to go back and forth just a few times:

For each of the following problems, balance the reaction by placing the appropriate coefficients in front of each compound.

PROBLEM 3.17.

$$P_2O_5 + H_2O \longrightarrow H_3PO_4$$

PROBLEM 3.18.

$$NH_4NO_3 \longrightarrow N_2O + H_2O$$

PROBLEM 3.19.

$$B_2O_3 + NaOH \longrightarrow Na_3BO_3 + H_2O$$

PROBLEM 3.20.

$$BCl_3 + H_2O \longrightarrow H_3BO_3 + HCl$$

PROBLEM 3.21.

$$(NH_2)_2CO + H_2O \longrightarrow CO_2 + NH_3$$

So far, we have seen two tools. In many cases, you will be able to use both tools to help you solve a problem. In cases like those shown above, you should use Tool #1 first. So, your general protocol for balancing reactions should be like this:

1. Write a list of the elements that you need to balance. Then look for any reagents that have only one type of element. You will balance these elements last.

2. Look for the compound with the largest number of different elements. Start with those elements. Begin with the elements that appear in the fewest compounds.

Now let's do a few problems where you can use both tools.

PROBLEM 3.22.

$$C_7H_8O_2 + O_2 \longrightarrow CO_2 + H_2O$$

PROBLEM 3.23.

$$As_4S_6 + O_2 \longrightarrow As_4O_6 + SO_2$$

PROBLEM 3.24.

$$S_2Cl_2 + NH_3 \longrightarrow S_4N_4 + S_8 + NH_4Cl$$

PROBLEM 3.25.

$$Ca_3(PO_4)_2 + SiO_2 + C \longrightarrow P_4 + CaSiO_3 + CO$$

Now we will do one last problem that is VERY difficult. If you find that you cannot balance the following reaction, no matter how hard you try, then you might want to take a look at Appendix A at the end of this book. The appendix gives an unorthodox approach to balancing reactions.

PROBLEM 3.26.

$$CaSO_4 + CH_4 + CO_2 \longrightarrow CaCO_3 + S + H_2O$$

3.3 INFORMATION TO IGNORE WHEN BALANCING REACTIONS

Most often, you will be given a reaction to balance where you are told the phase of each substance in the reaction (whether it is a solid, liquid, or gas). This information is usually italicized and placed in parentheses after each substance. For example:

$$N_2H_4 \ (l) + O_2 \ (g) \longrightarrow NO_2 \ (g) + H_2O \ (l)$$

This unbalanced reaction tells us that N_2H_4 is a *liquid*, O_2 is a *gas*, NO_2 is a *gas*, and H_2O is a *liquid*. The phases of these substances will be important later when you learn about thermodynamics. But for balancing reactions, this information is completely unnecessary. In fact, it often just confuses the issue, because you need to train your eye to ignore that information (for now).

Sometimes you will see *(aq)* after a substance. This means that the substance is dissolved in water (an aqueous solution). For now, you will ignore this also.

Let's do a few more problems. This time, we will be given the phase information. These problems are not any different than before. Just ignore the phase information.

Balance the following reactions:

PROBLEM 3.27.

$$NaCl \ (aq) + H_2O \ (l) \longrightarrow NaClO_3 \ (aq) + H_2 \ (g)$$

PROBLEM 3.28.

$$SO_3 \ (g) + H_2O \ (l) \longrightarrow H_2SO_4 \ (l)$$

PROBLEM 3.29.

$$SO_2 \ (g) + O_2 \ (g) \longrightarrow SO_3 \ (g)$$

PROBLEM 3.30.

$$N_2H_4 \ (l) + O_2 \ (g) \longrightarrow NO_2 \ (g) + H_2O \ (l)$$

PROBLEM 3.31.

$$(CN)_2 \ (g) + H_2O \ (l) \longrightarrow H_2C_2O_4 \ (aq) + NH_3 \ (g)$$

3.4   LIMITING REAGENTS

Now that we have seen how to balance reactions, we will turn our attention to the kinds of stoichiometric calculations that are based on balanced reactions. All of these calculations require that you first determine which reagent is the "limiting reagent". The concept of limiting reagents is not a difficult concept. The issue is that you will see problems that don't mention the words "limiting reagent", but nonetheless, the only way to solve these problems is to first calculate which reagent is the limiting reagent, and then base the rest of your calculations on that information. In this section, we will build slowly toward those kinds of problems by making sure that we know how to calculate which reagent is the limiting reagent.

Let's begin by defining the term "limiting reagent" using our Lego analogy from the previous sections. Consider the following instructions which show us how to build a simple Lego structure from one black piece and one white piece:

If we wanted to build a barrel of products, then we would need one barrel of black pieces and one barrel of white pieces. The ratio is 1:1. If we had two barrels of white pieces and two barrels of black pieces, then we could build two barrels of products. If we had 250 barrels of each kind of piece, then we could build 250 barrels of products. But what if we had unequal numbers of individual black and white pieces? Suppose we had ten barrels of black pieces and only three barrels of white pieces? How many products could we build? In this case, the "limiting reagent" is the white pieces. The quantity of white pieces determines how many products we can make. If we only have three barrels of white pieces, then we can only make three barrels of products.

In the case above, it was simple to determine which piece was the limiting reagent. We had fewer barrels of white pieces than black pieces, so it was simple.

But it was only simple because the ratio that we needed was 1:1. Consider the example below, where the ratio is not 1:1.

In this example, we need *two* white pieces for every *one* black piece. So, if we have *one* barrel of black pieces, we will need *two* barrels of white pieces. What if we have *five* barrels of black pieces and *eight* barrels of white pieces? Which one is the limiting reagent? It is true that we have more white pieces than black pieces, BUT the ratio is not 1:1 in this case. In order to use up all *five* barrels of black pieces, we would need *ten* barrels of white pieces. And we don't have that. We only have *eight* barrels of white pieces. So, the white pieces are the limiting reagent in this case, even though we have more white pieces than black pieces. From this example, we can clearly see that we absolutely MUST have a balanced reaction in order to determine which reagent is limiting.

It is important to be able to determine which reagent is limiting, because the limiting reagent always *determines* how much product we get. In the example we just saw, the white pieces were the limiting reagent, and therefore, we ignore the number of black pieces that we have, and we focus on how many white pieces we have. Since we only have eight barrels of white pieces, we will only be able to form four barrels of products (because we need *two* white pieces to make *one* product).

Now let's consider a slightly more complicated case. This is an example that we have seen at the beginning of this chapter. Consider this example again:

Notice the first term in the figure above. Recall that this example was the situation where the Lego warehouse didn't sell barrels of individual white pieces. They only sold barrels of white structures (where each white structure is composed of two white Lego pieces). The figure above indicates that we need *three* barrels of white structures and *two* barrels of black pieces in order to make *two* barrels of products. Suppose we have six barrels of white structures and six barrels of black pieces. Which one is the limiting reagent? The white structures or the black pieces?

The best way to think about this is to consider the white pieces and black pieces separately. And for each one, we ask ourselves how many products you could make. Let's start with white pieces. If we have six barrels of white structures, then we can look at the figure above and determine that we can make four barrels of products (because it takes three barrels of white structures to make two barrels of products). Now we do the same analysis for the black pieces. If we have six barrels of black pieces, then we can look at the figure above and determine that we can make six barrels of products. To summarize, we have enough white pieces to make four barrels of products, and we have enough black pieces to make six barrels of products. Therefore, the white pieces are the limiting reagent. In the end, we will only be able to make four barrels of products. If we had even more black pieces, it wouldn't help, because the limiting reagent determines how many products you can make.

The concept is exactly the same for chemical reactions. If you want to know how much product can be produced by a reaction, you need to know which reagent is the limiting reagent. AND you need to have a balanced reaction for two reasons: (1) so that you can determine which reagent is limiting, and (2) so that you can determine how many moles of product can be produced for every mole of limiting reagent.

That is why it is so important to know how to balance reactions. If you can't balance a reaction, then you won't be able to calculate how much product you can expect from a reaction. Let's see a specific example of this:

EXERCISE 3.32.   Consider the following unbalanced reaction

$$CO_2\ (g) + H_2\ (g) \longrightarrow CH_3OH\ (l) + H_2O\ (l)$$

If you are given *2* moles of CO_2 and *3* moles of H_2, determine how many moles of methanol (CH_3OH) will form.

Answer: You must begin by balancing the reaction:

$$CO_2\ (g) + \boldsymbol{3}\ H_2\ (g) \longrightarrow CH_3OH\ (l) + H_2O\ (l)$$

Now, you are ready to solve the problem. To determine which one is the limiting reagent, look at the "recipe" above, and ask how many moles of CH_3OH and H_2O can be produced from the materials that are given. Look at each of the starting reagents separately. Start with CO_2.

The problem tells you that you are given *2* moles of CO_2, so the recipe above indicates that you can potentially make *2* moles of CH_3OH and *2* moles of H_2O.

Now let's consider the other starting reagent. The problem tells you that you are given *3* moles of H_2, so the recipe above indicates that you can potentially make *1* mole of CH_3OH and *1* mole of H_2O. So, H_2 is the limiting reagent, and in this case, you can only form 1 mole of each of the products.

PROBLEM 3.33.   Consider the following ***unbalanced*** reaction:

$$C_2H_2\ (l) + H_2\ (g) \longrightarrow C_2H_6\ (g)$$

If you are given *6* moles of C_2H_2 and *6* moles of H_2, determine how many moles of product will form.

Answer: _____

PROBLEM 3.34. Consider the following *unbalanced* reaction:

$$N_2 \ (g) + H_2 \ (g) \longrightarrow NH_3 \ (g)$$

You are given *4* moles of N_2 and *9* moles of H_2. How many moles of product will form?

Answer: _____

PROBLEM 3.35. Consider the following *unbalanced* reaction:

$$CO \ (g) + O_2 \ (g) \longrightarrow CO_2 \ (g)$$

You are given *2* moles of CO and *2* moles of O_2. How many moles of product will form?

Answer: _____

PROBLEM 3.36. Consider the following *unbalanced* reaction:

$$Cl_2 \ (g) + O_2 \ (g) \longrightarrow Cl_2O \ (g)$$

You are given *4* moles of Cl_2 and *4* moles of O_2. How many moles of product will form?

Answer: _____

PROBLEM 3.37. H_2, O_2, and N_2 react with each other to form HNO_2. If you start with *2* moles of H_2, *2* moles of O_2, and *2* moles of N_2, how many moles of product will form?

Answer: _____

In all of the cases we have seen so far, we were given the amount of each reactant, measured in *moles*. But there is another way to ask the same question. Recall from Chapter 2 that we can interconvert mass and moles using the molar mass (which we can calculate whenever we are given a molecular formula). Therefore, you should expect to see problems where you are given the amount of each reactant, measured in *grams*.

To solve problems like this, you will first need to convert from grams into moles. Once you have done the conversion, then the problem is exactly like those above.

Let's see an example:

EXERCISE 3.38. Consider the following unbalanced reaction:

$$C_3H_8 \ (g) + O_2 \ (g) \longrightarrow CO_2 \ (g) + H_2O \ (l)$$

If you start with 10.02 g of propane (C_3H_8) and 16.5 g of O_2, how many moles of CO_2 will form?

Answer: In order to determine how many moles of product will form, you need to know how many moles of limiting reagent you have. So, you must first balance the reaction, because you cannot determine which reagent is limiting until you have done that. The balanced reaction is:

$$C_3H_8 \ (g) + 5 \ O_2 \ (g) \longrightarrow 3 \ CO_2 \ (g) + 4 \ H_2O \ (g)$$

Now that you have a balanced reaction, you will need to determine which reagent is limiting. But you are not given amounts in moles. Rather, you are given amounts in grams, so you must convert grams into moles using molar mass. Start by converting C_3H_8. The molar mass of C_3H_8 is: $3(12.01 \ g/mol) + 8(1.008 \ g/mol) = 44.09$ g/mol. To convert grams into moles, use the molar mass as a conversion factor:

$$10.02 \ g \ C_3H_8 \times \frac{1 \ mol \ C_3H_8}{44.09 \ g \ C_3H_8} = \underline{0.2273 \ mol \ C_3H_8}$$

Now, you need to do the same thing for the other starting reagent, O_2. The molar mass of O_2 is: $2(16.00 \ g/mol) = 32.00 \ g/mol$. Using this molar mass as a conversion factor, you can calculate how many moles of O_2 you are starting with:

$$16.5 \ g \ O_2 \times \frac{1 \ mol \ O_2}{32.00 \ g \ O_2} = \underline{0.516 \ mol \ O_2}$$

So far, you have calculated how many moles of C_3H_8 and how many moles of O_2 you start with. You have more moles of O_2 than C_3H_8, but that does **not necessarily** mean that C_3H_8 is the limiting reagent. To determine which reagent is limiting, you need to look at the balanced reaction. Consider each reagent separately, and ask yourself how much product would you get? The balanced reaction tells you that 1 mole of C_3H_8 can potentially produce 3 moles of CO_2. So, if you are starting with 0.2273 mol C_3H_8, then you have enough to make $3(0.2273 \ mol) = 0.6819 \ mol \ CO_2$.

Now consider how much O_2 you have to start with. The balanced reaction tells you that 5 moles of O_2 can potentially produce 3 moles of CO_2. So, if you are starting with 0.516 mol O_2, then you can have enough to make $(^3/_5)(0.516 \ mol) = 0.310 \ mol \ CO_2$.

When you compare these numbers, you'll see that you have enough C_3H_8 to form 0.6819 mol CO_2, but only enough O_2 to form 0.310 mol CO_2. So, you can conclude that O_2 is the limiting reagent.

The limiting reagent determines how much product you get. You can only make 0.310 mol CO_2.

Each of the following problems is exactly like the example we just saw. In each of these problems, you will be given an unbalanced reaction, and you will be asked to calculate how much product can be formed when given the amounts of starting reagents, measured in *grams*. To solve each of these problems, you will need to first balance the reaction, then convert grams into moles (by calculating the molar mass of each starting reagent and then using the molar mass as a conversion factor), then determine which reagent is limiting, and then determine how many moles of products are formed.

PROBLEM 3.39. Consider the following *unbalanced* reaction:

$$TiCl_4 \ (s) + H_2O \ (l) \longrightarrow TiO_2 \ (s) + HCl \ (aq)$$

How many moles of TiO_2 can be made from 50.00 g of $TiCl_4$ and 50.00 g of H_2O?

Answer: _____

PROBLEM 3.40. Consider the following *unbalanced* reaction:

$$P_4O_{10} \ (s) + H_2O \ (l) \longrightarrow H_3PO_4 \ (l)$$

How many moles of H_3PO_4 can be made from 150.00 g of P_4O_{10} and 15.00 g of H_2O?

Answer: _____

PROBLEM 3.41. Consider the following *unbalanced* reaction:

$$B_2H_6 \ (g) + H_2O \ (l) \longrightarrow H_3BO_3 \ (s) + H_2 \ (g)$$

How many moles of H_3BO_3 can be made from 34.72 g of B_2H_6 and 28.56 g of H_2O?

Answer: _____

PROBLEM 3.42. Consider the following *unbalanced* reaction:

$$AgNO_3 \ (aq) + Na_2SO_4 \ (aq) \longrightarrow Ag_2SO_4 \ (s) + NaNO_3 \ (aq)$$

How many moles of Ag_2SO_4 can be made from 14.07 g of $AgNO_3$ and 15.64 g of Na_2SO_4?

Answer: _____

Now, we are ready to take one more step. Suppose we have a problem exactly like the ones above, but we need to express our answer in grams of product, instead of moles of product. Once again, we know that we can always convert grams into moles or moles into grams when we have the molecular formula. We just use the molar mass as a conversion factor. So, we do everything that we just did in the previous problems, and then we end off with a conversion of moles of product into grams of product.

There are two reasons why problems like this seem tricky:

1. You are given the amount of starting reagents in grams, and you are asked to calculate how much product you expect in grams. The problem makes no mention of moles at all. But you have to realize that amounts of reagents are related to each other through "moles" (through the coefficients in the balanced reaction). So, even though the problem only talks about grams, you will have to convert everything into moles, and then back into grams when you are done.

2. There is no mention of the need to calculate which reagent is limiting, but you have to do so in order to be able to solve the problem.

So, the protocol goes like this: you will need to first balance the reaction, then convert grams into moles (by calculating the molar mass of each starting reagent and then using the molar mass as a conversion factor), then determine which reagent is limiting, then determine how many moles of products are formed, and then convert mole of product into grams of product using the molar mass of the product (which you can determine from the molecular formula). That is A LOT of steps. As far as problems go, this is pretty close to the largest number of steps that you will have to do in one specific problem.

EXERCISE 3.43. Consider the following unbalanced reaction:

$$C_3H_8\ (g) + O_2\ (g) \longrightarrow CO_2\ (g) + H_2O\ (l)$$

If you start with 10.02 g of propane (C_3H_8) and 16.5 g of O_2, what mass (in grams) of CO_2 will form?

Answer: This problem is exactly the same as the previous worked example, except here you are asked to report your answer in *grams* of product, rather than in *moles* of product. So, you do all of the steps that you did in the previous example, but then you must convert your answer from moles into grams.

Start by balancing the reaction; you get:

$$C_3H_8\ (g) + 5\ O_2\ (g) \longrightarrow 3\ CO_2\ (g) + 4\ H_2O\ (g)$$

Next, determine which reagent is limiting. To do this, you must convert grams into moles using molar mass. Start by converting C_3H_8. The molar mass of C_3H_8 is = 3(12.01 g/mol) + 8(1.008 g/mol) = 44.09 g/mol. Using the molar mass as a conversion factor, you get:

$$10.02\ \text{g}\ C_3H_8 \times \frac{1\ \text{mol}\ C_3H_8}{44.09\ \text{g}\ C_3H_8} = \underline{0.2273\ \text{mol}\ C_3H_8}$$

Now, do the same thing for O_2; you get:

$$16.5\ \text{g}\ O_2 \times \frac{1\ \text{mol}\ O_2}{32.00\ \text{g}\ O_2} = \underline{0.516\ \text{mol}\ O_2}$$

To determine which reagent is limiting, look at the balanced reaction, which tells you that you have enough C_3H_8 to potentially make 3(0.2273 mol) = **0.6819 mol CO_2**. And you have enough O_2 to potentially make ($^3/_5$)(0.516mol) = **0.310 mol CO_2**. Therefore, O_2 is the limiting reagent.

The limiting reagent determines how much product you get. So, you can only make 0.310 mol CO_2.

Finally, you need to convert your answer into grams of product. To do this, you need the molar mass of CO_2, which you can determine from the molecular formula. The molar mass of CO_2 is (12.01 g/mol) + 2(16.00 g/mol) = 44.01 g/mol.

Now use that as a conversion factor to get your answer:

$$0.310 \text{ mol CO}_2 \times \frac{44.01 \text{ g CO}_2}{1 \text{ mol CO}_2} = \underline{13.6 \text{ g CO}_2}$$

PROBLEM 3.44. Consider the following **unbalanced** reaction:

$$CH_4 \ (g) + Cl_2 \ (g) \longrightarrow CCl_4 \ (l) + HCl \ (g)$$

What mass (in grams) of CCl_4 can be made from 15.00 g of CH_4 and 45.00 g of Cl_2?

Answer: _____

PROBLEM 3.45. Consider the following **unbalanced** reaction:

$$Cr_2O_3 \ (s) + H_2S \ (g) \longrightarrow Cr_2S_3 \ (s) + H_2O \ (l)$$

What mass (in grams) of Cr_2S_3 can be made from 20.43 g of Cr_2O_3 and 100.0 g of H_2S?

Answer: _____

PROBLEM 3.46. Consider the following **unbalanced** reaction:

$$Fe(OH)_3 \ (s) + H_2SO_4 \ (aq) \longrightarrow Fe_2(SO_4)_3 \ (aq) + H_2O \ (l)$$

What mass (in grams) of $Fe_2(SO_4)_3$ can be made from 100.0 g of $Fe(OH)_3$ and 50.0 g of H_2SO_4?

Answer: _____

3.5 LIMITING REAGENT PROBLEMS CAN BE SOLVED IN A SINGLE CALCULATION

In all of the problems in the previous section, we developed a step-by-step strategy for calculating the mass of product that can be formed when you are given the mass of the reactants. However, there is a faster way to do these kinds of problems. It is possible to do all of the conversions *in a single calculation*. This will save valuable time on an exam, so let's take a look at how to do this. We will repeat the previous exercise (3.43), and we will show how it can be done more quickly. Let's see that question again:

EXERCISE 3.47. Consider the following unbalanced reaction:

$$C_3H_8 \ (g) + O_2 \ (g) \longrightarrow CO_2 \ (g) + H_2O \ (l)$$

If you start with 10.02 g of propane (C_3H_8) and 16.5 g of O_2, what mass (in grams) of CO_2 will form?

Answer: When we saw this problem in the previous section, we used the following strategy: We first balanced the reaction:

$$C_3H_8 \ (g) + 5 \ O_2 \ (g) \longrightarrow 3 \ CO_2 \ (g) + 4 \ H_2O \ (g)$$

Then we converted the mass of propane into moles, and we converted the mass of oxygen into moles. Then we used the balanced reaction to determine which one was the limiting reagent, which allowed us to determine how many moles of CO_2 would form. Then we converted that into mass of CO_2.

This time, we will do the same thing, but we will do all of the conversions in a single calculation. We will calculate the mass of product that we would expect from the given mass of *propane* (all in one calculation). Next, we will calculate the mass of product that we would expect from the given mass of *oxygen* (all in one calculation). Then, we will only need to ask which calculation generates the smallest mass of product. That will be our answer. Now let's see how we do this:

You are given 10.02 grams of propane. You can convert this into moles using the molar mass of propane. Then, you can convert that into moles of product using the following conversion factor (which you get from the balanced equation):

$$\frac{3 \ mol \ CO_2}{1 \ mol \ C_3H_8} = 1$$

Finally, you can convert the moles of product into mass, using the molar mass of the product. ***String all three of these conversions together into one calculation, and you get***:

$$10.02 \ g \ C_3H_8 \times \frac{1 \ mol \ C_3H_8}{44.09 \ g \ C_3H_8} \times \frac{3 \ mol \ CO_2}{1 \ mol \ C_3H_8} \times \frac{44.01 \ g \ CO_2}{1 \ mol \ CO_2} = 30.01 \ g \ CO_2$$

This tells you that 10.02 grams of propane could potentially produce ***30.01 g*** of CO_2.

Now, do the same exact procedure for the other reactant. You are given 16.5 grams of O_2. You can convert this into moles using the molar mass of O_2. Then, you can convert that into moles of product using the following conversion (which you get from the balanced equation):

$$\frac{3 \ mol \ CO_2}{5 \ mol \ O_2} = 1$$

Finally, you can convert the moles of product into mass, using the molar mass of the product. ***String all three of these conversions together into one calculation, and you get***:

$$16.5 \ g \ O_2 \times \frac{1 \ mol \ O_2}{32.00 \ g \ O_2} \times \frac{3 \ mol \ CO_2}{5 \ mol \ O_2} \times \frac{44.01 \ g \ CO_2}{1 \ mol \ CO_2} = 13.6 \ g \ CO_2$$

This tells you that 16.5 grams of O_2 could potentially produce ***13.6 g*** of CO_2.

When you compare the two calculations, you'll see that there is enough propane to produce 30.01 g of product and enough oxygen to produce 13.6 g of product. Therefore, oxygen is the limiting reagent, which means that you will only form 13.6 g of product.

Notice that you obtained the same answer as you did for this same problem in the previous section. But this time the calculations went much faster.

Now, you should try to do this procedure on problems similar to those you did earlier (Problems 3.44–3.46). But this time you should try to do all of the conversions in a single calculation (as you did in the previous exercise). You should find that this method will help you get your answer much faster.

PROBLEM 3.48.   Consider the following *unbalanced* reaction:

$$CH_4 \ (g) + Cl_2 \ (g) \longrightarrow CCl_4 \ (l) + HCl \ (g)$$

What mass (in grams) of CCl_4 can be made from 22.00 g of CH_4 and 40.00 g of Cl_2?

PROBLEM 3.49.   Consider the following *unbalanced* reaction:

$$Cr_2O_3 \ (s) + H_2S \ (g) \longrightarrow Cr_2S_3 \ (s) + H_2O \ (l)$$

What mass (in grams) of Cr_2S_3 can be made from 2.45 g of Cr_2O_3 and 1.00 g of H_2S?

PROBLEM 3.50.   Consider the following *unbalanced* reaction:

$$Fe(OH)_3 \ (s) + H_2SO_4 \ (aq) \longrightarrow Fe_2(SO_4)_3 \ (aq) + H_2O \ (l)$$

What mass (in grams) of $Fe_2(SO_4)_3$ can be made from 12.5 g of $Fe(OH)_3$ and 12.5 g of H_2SO_4?

Now that we have seen how to string all of the conversions into one calculation, we can appreciate that this method of problem solving can be applied to other, similar stoichiometry problems. For example, suppose we are given the mass of the product formed during a reaction, and we have to calculate the mass of reactants necessary to produce that much product. To solve this kind of problem, we use the same technique—we string all of the conversion factors together into one calculation. Let's see a specific example:

EXERCISE 3.51. Consider the following unbalanced reaction:

$$C_4H_{10} \; (l) + O_2 \; (g) \longrightarrow CO_2 \; (g) + H_2O \; (l)$$

What mass of butane (C_4H_{10}) is necessary in order to produce 10.0 g of CO_2?

Answer: Always begin by balancing the reaction:

$$2 \; C_4H_{10} \; (l) + 13 \; O_2 \; (g) \longrightarrow 8 \; CO_2 \; (g) + 10 \; H_2O \; (g)$$

You can solve this problem very much like you did in the previous set of problems by using a single calculation. But this time you will be working backwards. Start by converting the *mass* of product into *moles*. Then use the balanced equation to determine how many moles of *reactant* were necessary. And finally, convert *moles* of reactant into *mass* of reactant. Let's build up your calculation piece-by-piece.

You are given the *mass* of the product (10.0 g of CO_2), and you must first convert that into *moles* of CO_2. Use the molar mass as a conversion factor:

$$10.0 \text{ g } CO_2 \times \frac{1 \text{ mol } CO_2}{44.01 \text{ g } CO_2}$$

Then, you need to convert moles of CO_2 into moles of butane. The balanced reaction tells you how many moles of *butane* are necessary to produce this much CO_2. It takes 2 moles of butane to form 8 moles of CO_2. So, use this as a conversion factor, and place this conversion factor in your calculation above to look like this:

$$10.0 \text{ g } CO_2 \times \frac{1 \text{ mol } CO_2}{44.01 \text{ g } CO_2} \times \frac{2 \text{ mol } C_4H_{10}}{8 \text{ mol } CO_2}$$

Notice that this conversion factor has moles of butane in the numerator, and moles of CO_2 in the denominator. You were able to determine what to put in the numerator and what to put in the denominator because this is the only way that the units will cross off properly. The previous conversion factor has moles of CO_2 in the *numerator*, so the second conversion factor must have moles of CO_2 in the *denominator* (so that mol CO_2 will cross off). That is the power of the factor-label method. It allows you to verify that you are doing it right by checking the units.

Finally, you need to insert one last conversion factor into your *single* calculation. Convert *moles* of butane into *mass* of butane (using the molar mass of butane):

$$10.0 \text{ g CO}_2 \times \frac{1 \text{ mol CO}_2}{44.01 \text{ g CO}_2} \times \frac{2 \text{ mol C}_4\text{H}_{10}}{8 \text{ mol CO}_2} \times \frac{58.12 \text{ g C}_4\text{H}_{10}}{1 \text{ mol C}_4\text{H}_{10}}$$

Now you have all of your conversion factors in one calculation, and you can calculate your answer. When you do the calculation above, the answer is *3.30 g* of C_4H_{10}. This is the mass of butane that you would need in order to produce 10.0 g of CO_2.

Now let's try this procedure on some problems. You should try to do all of the conversions in a single calculation (as you did in the previous exercise).

PROBLEM 3.52.   Consider the following **unbalanced** reaction:

$$CH_4 \ (g) + Cl_2 \ (g) \longrightarrow CCl_4 \ (l) + HCl \ (g)$$

What mass of methane (CH_4) is necessary in order to produce 10.0 g of CCl_4?

PROBLEM 3.53.   Consider the following **unbalanced** reaction:

$$Cr_2O_3 \ (s) + H_2S \ (g) \longrightarrow Cr_2S_3 \ (s) + H_2O \ (l)$$

What mass of Cr_2O_3 is necessary in order to produce 10.0 g of Cr_2S_3?

And now, here are two more problems that are slightly different than the problems you have done. In these problems you will be given the mass of one reactant, and you will have to calculate how much mass of the other reactant would be necessary to completely react with the first reactant.

Although these problems are slightly different than the problems above, the procedure is exactly the same. For each of these problems, you will string three conversion factors together into one calculation that will give you the answer. In each case, the first and third conversion factors will be molar mass conversions (of one reactant and of the other reactant); and the middle conversion factor will be based on the balanced reaction (that tells you how many moles of one reactant are necessary to react with a mole of the other reactant). See if you can go through these problems on your own:

PROBLEM 3.54.   Consider the following *unbalanced* reaction:

$$Fe(OH)_3 \ (s) + H_2SO_4 \ (aq) \longrightarrow Fe_2(SO_4)_3 \ (aq) + H_2O \ (l)$$

What mass (in grams) of $Fe(OH)_3$ is necessary to completely react with 25.0 grams of H_2SO_4?

PROBLEM 3.55.   Consider the following *unbalanced* reaction:

$$C_2H_6 \ (g) + O_2 \ (g) \longrightarrow CO_2 \ (g) + H_2O \ (l)$$

What mass (in grams) of O_2 is necessary to completely react with 12.0 grams of ethane (C_2H_6)?

3.6   THEORETICAL AND ACTUAL YIELDS

In all of the problems in the previous section, we have just assumed that we will get the *maximum* amount of product that is possible. We have been determining which reagent is the limiting reagent, and then we used the balanced reaction to determine how much product we can *possibly* get. But, it rarely works this way in the lab. Usually, when we actually perform a reaction in a laboratory, we find that we get less than the **theoretical yield.** The theoretical yield is the amount of product that we think we should get. But the **actual yield** is usually lower. There are several reasons for this.

First of all, most reactions do not go to completion. Instead, an equilibrium is reached between reactants and products (you will learn more about this later). Also, we sometimes lose material when we transfer chemicals from one place to another. There are many other experimental factors that can also lead to a reduction in yield.

It is often helpful to know how much product we actually formed, *relative* to the theoretical yield. Did we get 99% of the amount of product that was possible to form, or did we only get 25% of the amount that was possible to form? That information is important when you are actually performing experiments. Obviously, you always want to get a good yield. This percentage is called the *percent yield* (% yield), and is defined as:

$$\% \text{ yield} = \frac{\text{actual yield}}{\text{theoretical yield}} \times 100\%$$

To make sure that this definition makes perfect sense to you, consider an example where we calculate that the theoretical yield is 100 grams of product. But when we actually run the reaction, we only get 90 grams of product. So, our reaction had a 90% yield.

It is pretty simple to make these kinds of calculations. So, you will rarely see a problem where you are given the theoretical yield and the actual yield, and then you are asked to calculate the % yield. A problem like that would be trivial. Instead, you will see problems like the ones in the previous section, where you have to make many calculations to determine what the theoretical yield is, and then the problem will tell you what the actual yield was so that you can calculate a % yield. Let's see an example.

EXERCISE 3.56. Consider the following **unbalanced** reaction, which shows the reaction between acetylene (C_2H_2) and bromine to form tetrabromoethane:

$$C_2H_2 \ (g) + Br_2 \ (l) \longrightarrow C_2H_2Br_4 \ (l)$$

In one experiment, 1.00 gram of C_2H_2 was placed in a sealed container with 25.00 grams of Br_2, and 6.85 grams of product were obtained. Calculate the % yield of this reaction.

Answer: This problem is very similar to the problems in the previous section, where you calculated the theoretical yield of the reaction. Here, you just need to add one extra step at the end, where you use the theoretical yield and the actual yield to calculate the % yield. So, your overall plan goes like this:

1. Begin by balancing the reaction.
2. Based on the balanced equation, calculate the mass of product that you would expect from the given mass of C_2H_2 (all in one calculation). Similarly, you

will calculate the mass of product that you would expect from the given mass of Br_2 (all in one calculation). Then, you only need to ask which calculation generates the smallest mass of product. That will tell you what mass of product you can expect as a theoretical yield.

3. Finally, use the calculated theoretical yield, together with the given information of actual yield, to calculate the % yield.

Begin by balancing the reaction; you get:

$$C_2H_2 \ (g) + 2 \ Br_2 \ (l) \longrightarrow C_2H_2Br_4 \ (l)$$

Next, calculate how much product could be produced from each of the reactants. Start with C_2H_2. You are given 1.00 gram of propane. You can use this to calculate the expected mass of the product, all in one calculation (as you did in the previous section). *String all three of these conversions together into one calculation, and you get*:

$$1.00 \ g \ C_2H_2 \times \frac{1 \ mol \ C_2H_2}{26.04 \ g \ C_2H_2} \times \frac{1 \ mol \ C_2H_2Br_4}{1 \ mol \ C_2H_2} \times \frac{185.84 \ g \ C_2H_2Br_4}{1 \ mol \ C_2H_2Br_4}$$

$$= 7.14 \ g \ C_2H_2Br_4$$

This tells you that 1.00 gram of C_2H_2 could potentially produce *7.14 g* of $C_2H_2Br_4$.

Now, do the same exact procedure for the other reactant. You are given 25.00 grams of Br_2. Use this to calculate the expected mass of the product, all in one calculation (as you did in the previous section). *String all three of these conversions together into one calculation, and you get*:

$$25.00 \ g \ Br_2 \times \frac{1 \ mol \ Br_2}{159.80 \ g \ Br_2} \times \frac{1 \ mol \ C_2H_2Br_4}{2 \ mol \ Br_2} \times \frac{185.84 \ g \ C_2H_2Br_4}{1 \ mol \ C_2H_2Br_4}$$

$$= 14.54 \ g \ C_2H_2Br_4$$

This tells you that 25.00 grams of Br_2 could potentially produce *14.54 g* of $C_2H_2Br_4$.

When you compare the two calculations, you can see that you have enough C_2H_2 to produce 7.14 g of product and enough Br_2 to produce 14.54 g of product. Therefore, C_2H_2 is the limiting reagent, which means that you will only form 7.14 g of product.

So, the theoretical yield of this reaction, based on the amounts of starting materials that you were given, is 7.14 g of product. Finally, calculate the % yield, based on an actual yield of 6.85 g:

$$\% \ yield = \frac{6.85 \ g \ C_2H_2Br_4}{7.14 \ g \ C_2H_2Br_4} \times 100\% = 95.9\%$$

PROBLEM 3.57. Consider the following reaction, which shows the production of ethanol from ethylene:

$$C_2H_4 \ (g) + H_2O \ (l) \longrightarrow C_2H_5OH \ (l)$$

In one experiment, 15.00 grams of ethylene (C_2H_4) were placed in a reaction vessel with 10.00 grams of H_2O. 23.25 grams of product were obtained. Calculate the % yield of this reaction.

PROBLEM 3.58. Consider the following reaction, which shows the production of acetic acid (CH_3CO_2H):

$$CH_3COCl \ (l) + H_2O \ (l) \longrightarrow CH_3CO_2H \ (l) + HCl \ (g)$$

In one experiment, 10.00 grams of acetyl chloride (CH_3COCl) was placed in a reaction vessel with 2.00 grams of H_2O. 6.20 grams of acetic acid (CH_3CO_2H) were obtained. Calculate the % yield of this reaction.

PROBLEM 3.59. Consider the following *unbalanced* reaction:

$$NH_3 \ (g) + CH_3Cl \ (g) \longrightarrow (CH_3)_4NCl \ (s) + HCl \ (g)$$

In one experiment, 1.73 grams of ammonia (NH_3) was placed in a reaction vessel with 23.5 grams of methyl chloride (CH_3Cl). 9.85 grams of tetramethyl ammonium chloride [$(CH_3)_4NCl$] was obtained. Calculate the % yield of this reaction.

PROBLEM 3.60. Consider the following balanced reaction:

$$C_3H_6O \ (l) + 2 \ CH_3OH \ (l) \longrightarrow C_5H_{12}O_2 \ (l) + H_2O \ (l)$$

In one experiment, 3.862 grams of acetone (C_3H_6O) was placed in a reaction vessel with 5.000 grams of methanol (CH_3OH). 6.552 grams of $C_5H_{12}O_2$ were obtained. Calculate the % yield of this reaction.

CHAPTER 4

THE IDEAL GAS LAW

In this chapter, we will learn tips and techniques for solving the most common types of problems that focus on the Ideal Gas Law ($PV = nRT$). Some of these problems can appear quite complex, even though they are actually very simple. You just need to train your eye to see the patterns, and you have to know what to look for.

Before we get started, we should recognize that there are two broad groups of problems: those describing a gas under a specific set of conditions, and those describing a gas undergoing a ***change*** of conditions. To illustrate the difference, imagine that you have a balloon filled with helium gas. We can make calculations on the gas inside the balloon, while there is no change taking place. Or, we can measure how the volume will expand if you heat the balloon. In the second case, our calculations revolve around a ***change*** in conditions. In order to quickly solve any Ideal Gas Law problem, you must train your eye to detect the difference between these two scenarios. You should always ask yourself: "Is there a change of conditions, or not?" You should be on the lookout for key phrases like these:

- the gas expands to a volume of . . .
- the temperature of the gas is raised to . . .
- the pressure on the gas is decreased to . . .

Whenever you read a problem with these (or similar) phrases, you should immediately recognize that the problem is referring to a change of conditions.

The first five sections of this chapter will focus on problems that pertain to a gas under a specific set of conditions (a gas that is ***not*** undergoing a change of conditions). The last three sections of the chapter are devoted to problems that pertain to a gas that ***is*** changing its conditions.

4.1 UNIT CONVERSIONS FOR THE IDEAL GAS LAW

In this chapter, we will see many different types of problems. In *every* problem, you must always make sure that your units are consistent ***before*** you do any calculations. In order for the equation ($PV = nRT$) to work properly, all of the units on the left side of the equation must be consistent with all of the units on the right side of the equation. This might seem hard to do, since the values for P, V, n, and T are all expressed using different units. That is where the value of R comes in handy. The value of R is the way that we get the units to be consistent.

R is just a constant: it is a number *with units attached to it*. The most common way to express R is like this:

$$R = 0.082057 \; \textbf{L atm / mol K}$$

Notice that the expression of R above has a number (0.082057), *and units* (L atm/ mol K). The units of R allow for the units in the equation to be consistent:

$$PV = nRT$$

$$(\text{atm})(\text{L}) = (\text{mol})\left(\frac{\text{L atm}}{\text{mol K}}\right)(\text{K})$$

Focus on the units on the right side (above): we see that *mol* can be crossed off (appearing in the numerator and denominator), and *K* can be crossed off as well. That leaves us with:

$$(\text{atm})(\text{L}) = (\cancel{\text{mol}})\left(\frac{\text{L atm}}{\cancel{\text{mol K}}}\right)(\text{K})$$

$$(\text{atm}) \; (\text{L}) = (\text{L}) \; (\text{atm})$$

So, we see that R is exactly what we need in order for the units on the right side to exactly match the units on the left side. But this only works because we used specific units for P, V, n, and T. We will soon see that these quantities can be measured with other units. If we had measured the values of P, V, and T using other units, then we would need to use a different expression for R. For example, here is another way of expressing R:

$$R = 8.3145 \; \text{J / mol K}$$

This expression of R is *not* a different value than the previous expression of R. Both expressions are describing the exact same quantity. As an analogy, we can say that a typical airplane travels at 600 miles per hour, or we can say that it travels at 880 feet per second. Either way, we are saying the same thing. In other words:

$$0.082057 \; \text{L atm / mol K} = 8.3145 \; \text{J / mol K}$$

And there are many other ways to express R as well. But for the most part, the two expressions shown above are the two most common expressions that you will use in this course. You will primarily use the first expression now (when we are solving gas law problems), and you will use the second expression for R in the second semester of this course (when you learn about thermodynamics)

So, for now, we will focus on using the first expression for R, which uses units of *L atm / mol K*. Since we are using this expression of R, we will need to have each term in the ideal gas law in the following units:

- *Pressure* must be measured in *atm*
- *Volume* must be measured in *Liters*
- *n* must be measured in *moles*
- *Temperature* must be measured in *Kelvin*

The most common mistake is to forget to convert units before solving a problem. For example, a problem will give you the temperature in *Celsius*, and you might forget to convert to *Kelvin*. Or the problem will give the pressure in *torr*, and you might forget to convert to *atm*. In this section, we will see all of the common conversions that you will have to do as a first step before solving any Ideal Gas Law problem. We will see that there are actually only a few common conversions that you will need to do. It is amazing how many problems in your textbook are simply testing you to see if you can recognize how to do the conversions first. Once you get in the habit of starting every problem by doing the proper conversions, you should be able to laugh at those problems (as if to say: "You can't trick me . . . I know how to do this . . .").

Let's start with *pressure*, since that is the first term in the equation ($\underline{P}V = nRT$). As mentioned earlier, we always want to express pressure in units of *atm*. These units make reference to the atmospheric pressure. Since atmospheric pressure changes with altitude and temperature, the *atm* unit (***standard atmosphere***) is the atmospheric pressure measured at sea level and 0° C. When a problem gives you the pressure expressed in units of *atm*, then you won't have to do any conversions.

But there are other units for pressure, and you will always need to convert those units into units of *atm*. The most common unit that you will see is mm Hg (millimeters of mercury). This unit is commonly called the *torr*. In your textbook, you will learn how pressure can be measured using a device called a manometer, in which measurements are made in mm of mercury. When you are given the pressure in these units, you will need to convert to units of *atm* using the following conversions:

$$1 \text{ atm} = 760 \text{ mm Hg}$$
$$1 \text{ atm} = 760 \text{ torr}$$

There is another common unit of pressure, based on the SI system. The SI unit for pressure comes from a combination of the SI base units ($\text{kg m}^{-1}\text{s}^{-2}$), and it is called the *Pascal*. It turns out that a Pascal is a very small unit of pressure, so you will see kiloPascals (kPa) more often than Pascals (Pa). When you have a problem that gives you pressure measured in *Pa* or *kPa*, you must convert into units of *atm* using the following conversions:

$$1 \text{ atm} = 101.325 \text{ kPa}$$
$$1 \text{ atm} = 101,325 \text{ Pa}$$

Finally, there is one more common unit for pressure, associated with the units we just saw. A *bar* is defined as 100 kPa. So, when you have a problem that expresses pressure in units of bars, you will need to use the following conversions:

$$1 \text{ atm} = 1.01325 \text{ bar}$$

Let's summarize the conversions we have seen, and then we can get some practice converting units. Here are the common conversions that you will need to convert pressure into units of *atm*:

$$1 \text{ atm} = 760 \text{ mm Hg}$$
$$1 \text{ atm} = 760 \text{ torr}$$

1 atm = 101.325 kPa

1 atm = 101,325 Pa

1 atm = 1.01325 bar

You might want to take a moment to notice the relationships between the various units above. Instead of thinking that there are five unrelated terms here, take a close look at the similarities. The first two units are related to each other (mm Hg and torr), and the last three are related to each other (Pa, kPa, and bar).

To use the conversions above, we will have to use the factor-label method (which we saw in Chapter 1). Let's see an example.

EXERCISE 4.1.   Convert 74.3 kPa into units of atm.

Answer: In order to convert kPa into atm, you will need to use the following conversion:

$$1 \text{ atm} = 101.325 \text{ kPa}$$

Use the factor-label method to get your answer:

$$74.3 \text{ kPa} \times \frac{1 \text{ atm}}{101.325 \text{ kPa}} = 0.733 \text{ atm}$$

If you don't remember the factor-label method, you should go back and review that section in Chapter 1.

For each of the following problems, convert into units of atm:

4.2.   1.342 bar = _____ atm

4.3.   1257 torr = _____ atm

4.4.   38 mm Hg = _____ atm

4.5. 2.53×10^5 Pa = _____ atm

4.6.   893 torr = _____ atm

4.7.   463 mm Hg = _____ atm

4.8.   151 kPa = _____ atm

4.9.   101 torr = _____ atm

So far, we have focused on pressure (the first term in the Ideal Gas Law). Now we move on to the second term: *volume* ($P\underline{V} = nRT$).

You must always convert volume into units of liters, and you will frequently see a problem where the volume is given in other units. Common units include mL, cm^3, dm^3, and m^3. It is fairly easy to convert from mL into L. We just use the following simple conversion:

$$1 \text{ L} = 1000 \text{ mL}$$

But it can be a bit trickier when you have to convert from cm^3, dm^3, or m^3. In these situations, it is helpful to remember that 1 mL = 1 cm^3. In fact, you should memorize this, because it will help you out many times throughout the course. You should always have this at your fingertips: 1 mL = 1 cm^3. Using this equality, we get:

$$1 \text{ L} = 1000 \text{ cm}^3$$

It can also easily be shown (using the factor-label method), that:

$$1 \text{ dm}^3 = 1 \text{ L}$$
$$\text{and}$$
$$1 \text{ m}^3 = 1000 \text{ L}$$

EXERCISE 4.10. Convert 2.50×10^4 cm^3 into units of liters.

Answer: In order to convert cm^3 into liters, you will need to use the following conversion:

$$1 \text{ L} = 1000 \text{ cm}^3$$

Use the factor-label method to get your answer:

$$2.50 \times 10^4 \text{ cm}^3 \times \frac{1 \text{ L}}{1000 \text{ cm}^3} = 25.0 \text{ L}$$

For each of the following problems, convert into units of liters:

4.11. 1342 cm^3 = _____ L   **4.12.** 1.37 dm^3 = _____ L

4.13. 896 ml = _____ L   **4.14.** 47.3 dm^3 = _____ L

4.15. 21 dm^3 = _____ L   **4.16.** 212 cm^3 = _____ L

4.17. 3.56×10^{-4} m^3 = _____ L

4.18. 54.5 cm^3 = _____ L

Now, we turn our attention to the third term in the Ideal Gas Law (PV = nRT). This term refers to the amount of gas that is present, measured in moles. In Chapter 2 (Section 2.5) we saw how to convert grams into moles using the molar mass as a conversion factor. This will be important here, because you will often see a problem where you are given the amount of gas in terms of grams (rather than moles). In these situations, you will need to convert into moles. Let's refresh our memory with the following exercise and problems:

EXERCISE 4.19. Convert 1.20×10^2 *grams* of CO_2 into *moles*.

Answer: This problem is asking you to convert grams into moles. To do this, you will need the molar mass, which the problem has not directly given you. But you do have the molecular formula, and that is all you need to calculate the molar mass: the molecular formula is CO_2. So, the molar mass is: $(12.01 \text{ g/mol}) + 2(16.00 \text{ g/mol}) = 44.01 \text{ g/mol}$.

Now, you have the formula you need to solve the problem:

$$1 \text{ mol of } CO_2 = 44.01 \text{ g}$$

Set up the following two conversion factors:

$$\frac{44.01 \text{ g } CO_2}{1 \text{ mol } CO_2} = 1 \quad \text{and} \quad \frac{1 \text{ mol } CO_2}{44.01 \text{ g } CO_2} = 1$$

Now, you just have to decide which one of these to use. In this problem, you are converting grams into moles. So, you will use the second conversion factor (above) in order that the units cancel off correctly to leave your answer in units of moles:

$$1.20 \times 10^2 \text{ g } CO_2 \times \frac{1 \text{ mol } CO_2}{44.01 \text{ g } CO_2} = 2.73 \text{ mol } CO_2$$

For each of the following problems, convert grams into moles:

4.20.   34.5 g of CH_4 = _____ mol

4.21.   52.3 g of C_2H_4 = _____ mol

4.22.   0.075 g of N_2 = _____ mol

4.23.   24 g of O_2 = _____ mol

4.24.   503 g of Cl_2 = _____ mol

And that brings us to the last term in the equation: *temperature* ($PV = nR\underline{T}$). In order to use the equation, we must always convert temperature into units of Kelvin. If we are given the temperature in Celsius, then we will need to use the following conversion:

$$K = {}^\circ C + 273.15$$

As a side note, we refer to Celsius units as *degrees*. As an example, we would say that the temperature outside is 20 *degrees* Celsius. But when we measure temperature in units of Kelvin, we don't use the word *degrees*. We would just say that the temperature is 293 Kelvin (without using the word *degrees*).

If we are given the temperature in Fahrenheit, then we will need to do two steps: first convert into Celsius, and then convert into Kelvin. For the first step (converting Fahrenheit into Celsius), we will need to use the following conversion:

$${}^\circ C = {}^5/_9 ({}^\circ F - 32)$$

EXERCISE 4.25.   Convert 23° F into Kelvin.

Answer: Begin by converting °F into °C. To do this use the conversion above and you will find that:

$$°C = \frac{5}{9} \, (°F - 32)$$
$$= \frac{5}{9} \, (23 - 32)$$
$$= \frac{5}{9} \, (-9)$$
$$= -5° \, C$$

Next, convert Celsius into Kelvin. To do this, add 273.15 to the temperature in Celsius, and you get:

$$-5 + 273.15 = \textbf{268 Kelvin}$$

Notice that your answer is not 268.15 (don't forget about significant figures). If you need a review of significant figures when adding or subtracting two numbers, go back to Chapter 1 (Section 1.1)

For each of the following problems, convert the temperature into Kelvin:

4.26. 100° C = _____K   **4.27.** 32° F = _____K

4.28. 25.7° C = _____K   **4.29.** 98.6° F = _____K

4.30. −15° C = _____K   **4.31.** 212° F = _____K

Before you move on to do some "real" problems in the next section, let's just make sure that you have mastered all of the conversions. Let's consider a problem where you have to convert all of the terms before you can begin to solve the problem. Consider the following example:

EXERCISE 4.32. *2.80 grams of CO_2 gas, occupying a volume of 54.7 cm^3 under a fixed pressure of 658 torr. ....* This is the way a typical Ideal Gas Law problem will start off. In order to make sure that you are ready to start learning techniques for solving these problems, let's make sure that you can do the conversions that you will need. Look at the terms above, and make any conversions that will be necessary in order to use the Ideal Gas Law.

Answer: You will need to convert grams into moles (using the molar mass of CO_2, which you can easily determine from the molecular formula), cm^3 into L, and torr into atm. These conversions are shown below:

$$2.80 \text{ g } CO_2 \times \frac{1 \text{ mol } CO_2}{44.01 \text{ g } CO_2} = 6.35 \times 10^{-2} \text{ mol } CO_2$$

$$54.7 \text{ cm}^3 \times \frac{1 \text{ L}}{1000 \text{ cm}^3} = 5.47 \times 10^{-2} \text{ L}$$

$$658 \text{ torr} \times \frac{1 \text{ atm}}{760 \text{ torr}} = 0.866 \text{ atm}$$

Each of the following problems represents a sample opening statement for Ideal Gas Law problems. For each of the opening statements below, make any necessary conversions.

PROBLEM 4.33. 5.5 grams of Cl_2 gas, occupying a volume of 2.7 dm^3 under a fixed pressure of 215 kPa. . . .

PROBLEM 4.34. 5.5 grams of O_2 gas, occupying a volume of 3.7×10^{-3} m^3 under a fixed temperature of 20° C. . . .

PROBLEM 4.35. 25.0 grams of C_2H_6 gas, occupying a volume of 50.0 cm^3 under a fixed pressure of 760 torr. . . .

There is one last comment that we need to make about conversions. We have to consider significant figures once again. When we converted −5° C, we got 268 K. We didn't report our answer as 268.15 K, because of the rules of significant figures. BUT, when we are planning to use this value and plug it into the Ideal Gas Law, the rules of significant figures tell us that we have to carry through the extra figures, and then only report *our final answer* with the correct number of significant figures. So, for all of the problems in the rest of this chapter, we will carry through some extra figures until the end of the problem (we will use 268.15 K instead of 268 K). And then, when we report our final answer, we will make sure that the answer is reported with the correct number of significant figures.

Similarly, when we use R in our calculations, we will have to decide how many significant figures to use in our value for R. In the case of temperature (above), it was simple—we just carried the two extra digits along for the ride (268.15 instead of 268). But in the case of R, how many digits do we need to carry? The value for R has *a lot* of digits. Do we really need to use 0.0820578 . . . ?

It certainly can't hurt to type extra significant figures into your calculator, but in most cases, this will be a waste of time. In most problems, you will be given data that is measured to two or three significant figures. Therefore, we can get away with using three or four significant figures for R (0.0821 or 0.08206). That way, we know that the precision of our result will be limited by the measured values, and not by the value of R that we used.

4.2 SIMPLE CALCULATIONS USING THE IDEAL GAS LAW

Take a close look at the Ideal Gas Law:

$$PV = nRT$$

R is constant, so there are only four variables in this equation (P, V, n, and T). If the problem gives you the values for any three of these terms, you will always be able to calculate the value of the fourth term.

These problems are perhaps the easiest types of Ideal Gas Law problems. You just need to remember to apply the necessary conversion factors (the stuff we saw in the previous section), and then you just plug and chug into the equation above. But even these "easy" problems can get disguised as tricky problems by forcing you to use *special* conversion factors. Once you train your eye to recognize these special conversion factors, you will see that the "tricky" problems are really just "easy" problems in disguise.

In this section, you will focus on the "easy" problems, without the special conversion factors, just to make sure that you get into the habit of calculating the fourth variable when you are given the values of the other three variables. Then, in the next section, you will see how these problems can appear to be tricky with special conversion factors.

EXERCISE 4.36.   0.25 moles of a gas occupies a volume of 545 cm^3 at a temperature of 25° C. Calculate the pressure of the gas.

Answer: The best way to start is to write down the information that you have been given:

n = 0.25 moles

V = 545 cm^3

T = 25° C

P = _____

By writing out the information, you can clearly see that you have been given three out of the four variables, and just need to plug them into the Ideal Gas Law in order to calculate the fourth variable. But remember that you must always do your conversions first.

The three values that you have are n, V, and T. The value of n is already in the correct units (moles), so you can move on to the next variable. Volume is given in cm^3, and it needs to be in liters, so use the factor-label method to make the conversion:

$$V = 545 \text{ cm}^3 \times \frac{1 \text{ L}}{1000 \text{ cm}^3} = 0.545 \text{ L}$$

Next, move on to the third variable that was given. The temperature was given in Celsius, so you need to convert to Kelvin:

$$T = 25° \text{ C} = (25 + 273.15) \text{ K} = 298.15 \text{ K}$$

Notice that some insignificant figures have been kept in the value for temperature (the value of 298.15 K is used, rather than 298 K). This is done because this value is not your final answer. Rather, it is an "intermediate answer" that will now be plugged in to the Ideal Gas Law. So, carry the insignificant figures along for the ride until the end of the calculation—and then only report the significant figures as your final answer.

Now, you can rewrite the three pieces of information that you have. Rewrite them in the units that you will need in order to use the Ideal Gas Law:

n = 0.25 moles

V = 0.545 L

T = 298.15 K

P = _____

Now you are ready to plug and chug:

$$PV = nRT$$

$$P = \frac{nRT}{V}$$

$$= \frac{(0.25 \text{ mol})(0.0821 \text{ L atm mol}^{-1} \text{ K}^{-1})(298.15 \text{ K})}{(0.545 \text{ L})}$$

$$= 11 \text{ atm}$$

Notice that your final answer has only two significant figures because the value for n (n = 0.25 mol) has only two significant figures. Problems like this are not difficult. But before you tackle the special conversion factors, you must be absolutely sure that you know how to do these straightforward problems. Let's just do a few practice problems, and then we will move on.

PROBLEM 4.37. 1.00 mole of a gas occupies a volume of 1.00 m³ at a temperature of 25° C. Calculate the pressure of the gas.

PROBLEM 4.38. 2.00 moles of a gas occupy a volume of 1.00 m³ at a pressure of 900 torr. Calculate the temperature of the gas.

PROBLEM 4.39. 2.50 moles of a gas are subjected to a temperature of 150° C and 500 kPa. Calculate the volume occupied by the gas.

PROBLEM 4.40.   A canister with a volume of 1.25 m^3 contains a gas at a temperature of 15° C and a pressure of 450 torr. Calculate the number of moles of gas present in the canister.

4.3 USING SPECIAL CONVERSION FACTORS

Now, we will see how the "easy" problems can be disguised as tricky problems through the use of special conversion factors. So far, we have seen two special conversion factors: density and molar mass. In Chapter 1, we learned that *density* is a measure of *mass per unit volume*, and we learned how to use density as a special conversion factor—to convert from mass to volume, or to convert from volume to mass (Section 1.6). Then in Chapter 2, we learned that molar mass is a measure of *mass per mole*, and we learned how to use molar mass as a special conversion factor—to convert from mass to moles, or to convert from moles to mass (Section 2.5). Now, these two special conversion factors are going to play a very important role.

To see how these conversion factors are delicately intertwined with the Ideal Gas Law, let's use diagrams to show it visually. We will start with molar mass.

When you know the molar mass of a compound (which you can calculate from the molecular formula—see Section 2.4), then you can easily convert grams into moles, or moles into grams. We can visually connect this to the Ideal Gas Law as shown here:

$$P \quad V \quad = \quad \boxed{\begin{array}{c} n \\[4pt] \updownarrow \text{\scriptsize molar mass} \\[4pt] \text{mass} \\ \text{(g)} \end{array}} \quad R \quad T$$

Carefully look at the contents of the box above. There are *three* quantities: n, molar mass, and mass (in grams). If we are given any two of these three quantities, then we can calculate the third. Let's summarize this information:

- If we are given the **mass** (in grams) and the **molar mass**, then we can calculate **n**.

- If we are given *n* and the *molar mass*, then we can calculate the *mass*.
- If we are given the *mass* and *n*, then we can calculate the *molar mass*.

Now, let's do a similar analysis with density. When you know the density of a compound, then you can convert grams into volume, or volume into grams. We can visually connect this to the Ideal Gas Law as shown here:

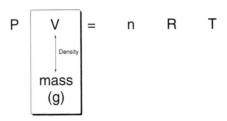

Once again, carefully look at the contents of the box above. There are *three* quantities: V, density, and mass (in grams). If you are given any two of these three quantities, then you can calculate the third. Let's summarize this information:

- If we are given the *mass* and the *density*, then we can calculate *V*.
- If we are given *V* and the *density,* then we can calculate the *mass*.
- If we are given the *mass* and *V*, then we can calculate the *density*.

Now we are ready to combine all of this information into one visual picture that shows all of these relationships:

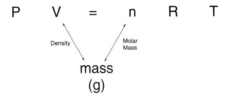

This chart shows everything we have seen, and it demonstrates the real usefulness of the Ideal Gas Law. If you want to be an expert at the "tricky" problems, then you will need to master this diagram. So, let's take a closer look.

If you know any *three* of the four variables circled below (remember that R is just a constant), then you can calculate the fourth:

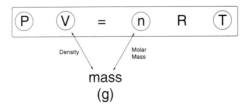

Similarly, if you know any *two* of the three variables circled below, then you can calculate the third:

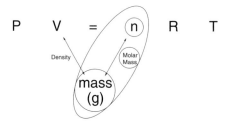

And finally, if you know any *two* of the three variables circled below, then you can calculate the third:

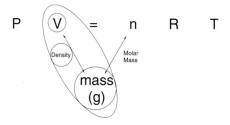

Now we have created a closed loop that allows us to solve some problems that might seem tricky at first. But with a bit of practice, they are actually quite easy. Let's see a specific example:

EXERCISE 4.41. Imagine that you are working on a problem that gives you the information circled below:

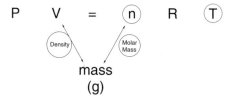

How would you calculate the pressure?

Answer: Begin by looking at the top equation (PV = nRT). You can see that you are given two of the four variables. You are given n and T, but you are not given P and V, so you cannot just plug and chug. You would also need V in order to calculate P.

Next look at the density relationship, and you'll see that you are stuck again. Of the three variables, you only have one (density), and you are missing volume and mass:

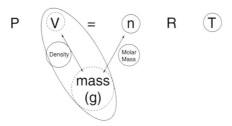

So, finally, turn to the molar mass relationship. And here, you are in luck. You have two of three variables, so you can calculate the third:

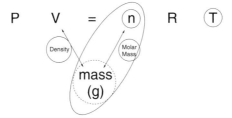

You are given n and molar mass, so you can calculate the mass.

Here comes the domino effect: Once you have calculated the mass, you can easily calculate the volume because you have two of the three variables you need:

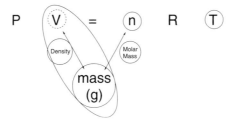

And once you have calculated the volume, you can calculate the pressure, because you have three of the four variables you need:

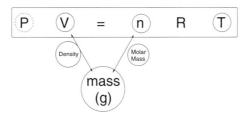

Let's see one more example:

EXERCISE 4.42. Imagine that you are working on a problem that gives you the information circled below:

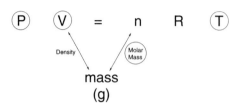

How would you calculate the density of the gas?

Answer: Start with the top equation ($PV = nRT$), and you'll see that you have three of the four variables. You are given P, V, and T. So, you can easily calculate n.

Once you have calculated n, you can use the molar mass to calculate the mass, because you have two of three variables:

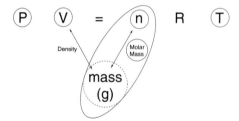

And once you know the mass, you can use the mass and the volume to calculate the density.

Notice how you use the domino effect, going around the closed loop of equations, to calculate your answer.

When the problems above are given as word problems, without any diagrams, they will appear to be *very tricky* problems. But, once you have seen how the two special conversion factors form a closed loop around the Ideal Gas Law, you can begin to appreciate that these problems are actually fairly easy to do. You just have to "see" it. In a few moments, we will actually do some real problems. For now, let's make sure that you can work through the logic, just like we did in the previous two exercises.

PROBLEM 4.43. Imagine that you are working on a problem that gives you the information circled below. How would you calculate T?

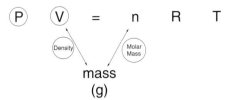

Answer: _____

PROBLEM 4.44. Imagine that you are working on a problem that gives you the information circled below. How would you calculate the molar mass?

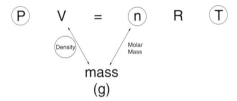

Answer: _____

PROBLEM 4.45. Imagine that you are working on a problem that gives you the information circled below. How would you calculate the density of the gas?

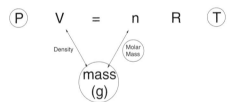

Answer: _____

PROBLEM 4.46. Imagine that you are working on a problem that gives you the information circled below. How would you calculate the pressure of the gas?

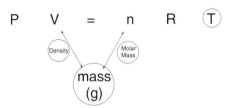

Answer: _____

Now that you know how this works, you are ready to start tackling some real problems. We will begin every problem by drawing a diagram (like the ones above) as a tool for devising a strategy to solve the problem. The goal is (hopefully) that you will see that these problems are not so tricky after all. In fact, they can actually be quite *fun* (believe it or not) when you know how to approach them. Let's see an example:

EXERCISE 4.47.   4.2 g of an unknown gas occupies a volume of 45 cm³ at a temperature of 25° C. The molar mass of the gas is 26.04 g/mol. Calculate the pressure of the gas.

Answer: Begin by drawing a diagram so that you can design a strategy for solving the problem.

Drawing the diagram is actually quite easy—you don't really have to memorize the diagram. Draw it like this: Start with the Ideal Gas Law:

$$P V = n R T$$

Then place *mass* at the bottom, underneath the equals sign:

$$P V = n R T$$
$$\text{mass}$$
$$\text{(g)}$$

Then, finish off by asking yourself these questions:

1. How is mass related to V?—*density.*

2. And how is mass related to n?—*molar mass.*

And now you have your diagram:

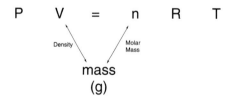

Now, circle what you are given. The problem gives you values for mass, V, T, and molar mass. Use your diagram to visualize the information you are given. Circle the given values:

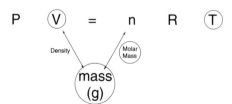

When you inspect the diagram above, the strategy becomes clear. You can use mass and molar mass to calculate n. Then, you can use n (together with V and T) to calculate P. Now that you have your strategy, use it to solve the problem.

You can start by using mass and molar mass to calculate n. In other words, use molar mass as a special conversion factor to convert grams into moles:

$$4.2 \text{ g} \times \frac{1 \text{ mol}}{26.04 \text{ g}} = 0.1613 \text{ mol}$$

You have calculated that n = 0.1613 mol. Notice once again that this number has too many significant figures. Technically, you should report this answer as n = 0.16 mol. BUT, you are not done with your calculations. You will need to plug this value in to the Ideal Gas Law. So, carry along a couple of extra digits, even though they are not significant. In the end, you will report your final answer with only two significant figures.

Now you have values for n, V, and T. So, you can use the Ideal Gas Law to solve for P. But remember, you must always convert the given information into the necessary units. The volume is given in cm^3, so you must convert that into L:

$$45 \text{ cm}^3 \times \frac{1 \text{ L}}{1000 \text{ cm}^3} = 4.5 \times 10^{-2} \text{ L}$$

Next, convert °C into Kelvin:

$$25° \text{ C} = (25 + 273.15) \text{ K}$$
$$= 298.15 \text{ K}$$

Now, you have the following information:

n = 0.1613 mol (extra figures carried along for the ride)

V = 4.5×10^{-2} L

T = 298.15 K (extra figures carried along for the ride)

P = _____

To finish off the problem, just plug in all of the values (with the proper units) into the Ideal Gas Law, and you get your answer:

$$PV = nRT$$

$$P = \frac{nRT}{V}$$

$$= \frac{(0.1613 \text{ mol})(0.0821 \text{ L atm mol}^{-1} \text{ K}^{-1})(295.15 \text{ K})}{(4.5 \times 10^{-2} \text{ L})}$$

$$= 87 \text{ atm}$$

Let's see one more example, and then you can try to do some problems yourself.

EXERCISE 4.48. A canister containing 1.00 mole of a gas is subjected to a pressure of 740 torr. The molar mass of the gas is 28.02 g/mol. At what temperature will that gas have a density of 1.25 g/L?

Answer: Begin by drawing a diagram and circling the given information. You are given n, P, molar mass, and density:

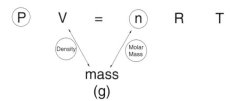

When you inspect the diagram above, the strategy becomes clear. You can use n and the molar mass to calculate the mass. Then, you can use the mass and the density to calculate V. And once you have V, you will be able to calculate T (using P, V, and n). Now that you have your strategy, use it to solve the problem.

Start by using n and molar mass to calculate grams. In other words, use molar mass as a special conversion factor to convert moles into grams:

$$1.00 \text{ mol} \times \frac{28.02 \text{ g}}{1 \text{ mol}} = 28.02 \text{ g}$$

(you are carrying along an extra digit that is not significant).

Now use the domino effect: Using the mass you just calculated (together with the density), you can calculate the volume. In other words, you are using density as a special conversion factor to convert mass into volume:

$$28.0 \text{ g} \times \frac{1 \text{ L}}{1.25 \text{ g}} = 22.4 \text{ L}$$

Now that you calculated V, you are ready to use P, V, and n to solve for T. But remember, you must always convert the given information into the necessary units. The pressure is given in torr, so you must convert to atm:

$$740 \text{ torr} \times \frac{1 \text{ atm}}{760 \text{ torr}} = 0.9737 \text{ atm}$$

(once again, you are carrying along an extra digit that is not significant).

The volume is given in L, so you don't need to do any unit conversions. And n is given in moles, so you are ready to finish the problem. In summary:

$P = 0.9737$ atm

$V = 22.4$ L

$n = 1.00$ mol

$T = $ _____

Now just plug in all of the values (with the proper units) into the Ideal Gas Law, and you get your answer:

$$PV = nRT$$
$$T = \frac{PV}{nR}$$
$$= \frac{(0.9737 \text{ atm})(22.4 \text{ L})}{(1.00 \text{ mol})(0.0821 \text{ L atm mol}^{-1} \text{ K}^{-1})}$$
$$= 266 \text{ K}$$

Now try to do each of the following problems. For each problem, use the space provided to develop your strategy, do your conversions, and calculate your answer.

PROBLEM 4.49.   A canister contains 2.05 mol of a gas that is known to have a molar mass of 32.00 g/mol. The pressure of the gas is 850 torr and the temperature of the gas is 50° C. Calculate the density of the gas under these conditions.

PROBLEM 4.50.   10.5 g of an unknown gas occupies a volume of 2.7 L at a pressure of 700 torr. The molar mass of the gas is 36.40 g/mol. Calculate the temperature of the gas.

PROBLEM 4.51. A 750 cm³ canister contains a gas at 100° C and a pressure of 780 torr. The molar mass of the gas is 28.02 g/mol. Calculate the density of the gas under these conditions.

PROBLEM 4.52. A 1.00 m³ canister contains a gas at a temperature of 72.5° F. The molar mass of the gas is known to be 32.00 g/mol, and the density (under these conditions) is measured to be 7.50 g/L. Calculate the pressure of the gas in the canister.

4.4 EXTENSIVE AND INTENSIVE PROPERTIES

So far, we have seen how helpful it can be to draw a diagram and circle the values that are given in a problem. In this section, we will take one more step toward intimately understanding these diagrams. To do so, let's consider a specific example. Suppose you have a problem where you are given P, T, and the molar mass:

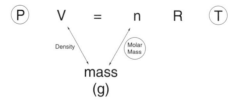

In this problem, you are then asked to calculate the density. If you use the method you learned in the previous section, you would conclude that it is **not** possible to calculate the density. It seems as though you just don't have enough information. But, there is a special trick that you can use to solve this problem. To understand how to use the trick, and more important, *when you are allowed* to use the trick, we need to understand that all of the properties (shown on the diagram above) can fall into one of two categories: *extensive* properties and *intensive* properties. Once we have a better understanding of the difference between these two categories, we will be in a better position to understand how we can solve the problem above.

Imagine that you have a balloon filled with helium gas. You decide that you would rather have two smaller balloons than one large balloon, so you transfer all of the helium into two smaller balloons. These balloons are now each half of the size of the original balloon. Now let's focus on *one* of these small balloons, and let's compare it to the original balloon. Some of the properties have changed, but others have not. The *volume* of one small balloon will certainly be less than the original balloon. Similarly, the number of *moles* in a small balloon will also be less than the number of moles that were in the original balloon. The same is true with *mass*. Each of these three properties is different when we compare one of the small balloons with the original balloon.

But some properties did not change. The *temperature* of the gas in a small balloon is the same as it was in the large balloon. Similarly, the *pressure* of the gas in a small balloon is also the same as the pressure was in the large balloon. *Density* and *molar mass* are two other properties that remain constant when we compare a small balloon with the original balloon.

So, we see that there are two kinds of properties: those that are dependent on the amount of gas, and those that are not dependent on the amount of gas. Each of these categories has a special name. Properties that **are** dependent on the amount of gas are called *extensive*, and properties that **are not** dependent on the amount of gas are called *intensive*.

Now let's take a look at our diagram again, and you should notice that all of the extensive properties are concentrated together in the center of the diagram:

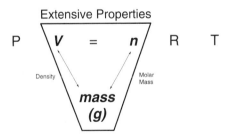

V, n, and mass are all *extensive* properties.

All of the other properties on the diagram above are *intensive*. P, T, density, and molar mass are all intensive properties. These properties do **not** depend on the amount of gas present. So, imagine that we have some gas in a balloon, and we know the pressure, the temperature, the molar mass, and the density (in other words, we know all of the intensive properties). In a situation like that, then the three extensive properties (*V*, *n*, and ***mass***, shown in the center of the figure above) are just describing how much gas we have in the balloon.

Now that we have seen our diagram in a new light, let's consider the example that seemed impossible to do:

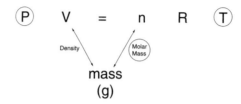

Let's think about what we are given in this problem. We are given P, T, and molar mass. All of these properties are intensive—they are ***not*** dependent on the amount of gas present. The problem asks us to calculate the density, which is also an intensive property. So, the problem can be rephrased like this: given the following three intensive properties, calculate the fourth intensive property. And here comes the trick: in a problem like this, it doesn't matter how much gas you have. You could be talking about a small helium balloon, or a huge balloon that is the size of a house. It doesn't matter. So, we can just pick any value of n, and we should be able to get our answer. Now we'll see how this works.

If we are going to choose a value for n, then let's choose a simple value, like n = 1 mole. Now, let's analyze our diagram again, but this time we have a value for n, so we place a circle around n in the diagram:

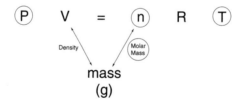

When analyzing this diagram, you should be able to see that you can now calculate the density. We use P, n, and T to calculate V. We use n and molar mass to calculate mass. And finally, we use mass and volume to calculate the density.

If we had chosen a value of n = 2 moles, then we would have gotten the same answer for density. Let's try to understand why. If we had 2 moles instead of 1, then the *mass* would be twice as large ***and*** the *volume* would also be twice as large. When you calculate density (dividing the mass by the volume), the answer you get will be the same as if you chose n = 1 mole. In fact, you can choose n = 100 moles, and you will still get the same answer!

Let's see a specific example of this:

EXERCISE 4.53.   At 100° C and a pressure of 720 torr, calculate the density of methane gas (CH_4).

Answer: Begin by analyzing what you are given. You are given the temperature. You are given the molecular formula, so you can easily calculate the molar mass. And, you are given the pressure. Now, draw your diagram to get a visual sense of what information you have:

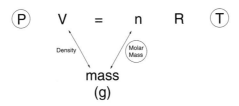

As you inspect this diagram, you notice that you are stuck. It seems as though you don't have enough information to solve for the density. BUT WAIT! You were only given intensive properties, and you are asked to calculate an intensive property. You should be able to do that by just using the value n = 1.

If you have a value for n, then you can calculate V, and you can calculate the mass. Then use those two values to calculate the density.

Now you have your strategy, but remember: before you do any calculations, you always have to convert your units:

$$720 \text{ torr} \times \frac{1 \text{ atm}}{760 \text{ torr}} = 0.9474 \text{ atm}$$

$$100° \text{ C} = (100 + 273.15) \text{ K}$$
$$= 373.15 \text{ K}$$

Now you are ready to implement your strategy. To calculate V, use the Ideal Gas Law:

$$PV = nRT$$
$$V = \frac{nRT}{P}$$
$$= \frac{(1 \text{ mol})(0.082057 \text{ L atm mol}^{-1} \text{ K}^{-1})(373.15 \text{ K})}{(0.9474 \text{ atm})}$$
$$= 32.333 \text{ L}$$

Notice that you used extra significant figures for R and T. And the answer you obtained (32.333 L) was written with too many significant figures (P was measured with three significant figures, so your answer should not have more than three significant figures). Carry a few extra digits along for the ride until the end of the calculation. Then you can make sure that your final answer has the correct number of significant figures.

Next, calculate the mass. You have n (which you defined as n = 1), and you need the molar mass, which you can easily determine from the molecular formula:

$$CH_4 = (12.01 \text{ g/mol}) + 4(1.008 \text{ g/mol}) = 16.04 \text{ g/mol}$$

Now, use that as a conversion factor to convert moles into mass:

$$1 \text{ mol} \times \frac{16.04 \text{ g}}{1 \text{ mol}} = 16.04 \text{ g}$$

You're almost done. You have now calculated V and mass, so you can use those values to calculate the density:

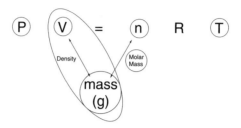

To do the calculation, recall what density means. It is a measure of the amount of mass in a certain volume. For gases, we usually measure density in g/L. So,

$$\text{Density} = \frac{\text{grams}}{\text{volume}} = \frac{16.04 \text{ g}}{32.333 \text{ L}} = 0.496 \text{ g/L}$$

If you try to do this problem again (from the beginning), using n = 2 instead of n = 1, you will find that you get the same answer. (Notice that your final answer was reported with three significant figures.)

Now try to do each of the following problems. For each problem, use the space provided to develop your strategy, do your conversions, and calculate your answer.

PROBLEM 4.54. Calculate the density of carbon dioxide gas (CO_2) at a temperature of 50° C and a pressure of 850 torr.

PROBLEM 4.55. Calculate the density of oxygen (O_2) at a temperature of 30° C and a pressure of 760 torr.

The trick that you have just learned applies to any situation where you are given three intensive properties and asked to calculate the fourth. Recall which properties on this diagram are intensive:

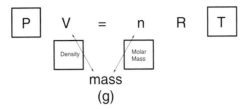

If you are given any three of these of these four quantities, you will always be able to calculate the fourth quantity by using $n = 1$ mol. Let's do some problems like that:

PROBLEM 4.56. A gas with a molar mass of 28.02 g/mol is cooled to a temperature of 0° C. The density of the gas is measured to be 1.03 g/L. What is the pressure of the gas?

PROBLEM 4.57. A gas with a molar mass of 28.02 g/mol is subjected to a pressure of 4.50 bar. The density of the gas is measured to be 3.56 g/L. What is the temperature of the gas?

PROBLEM 4.58.   An unknown gas is subjected to a pressure of 980 torr and a temperature of 10° C. The density of the gas is measured to be 0.89 g/L. What is the molar mass of the gas?

4.5   MOLAR MASS AS A BRIDGE

In the previous sections, we learned how to strategize using the following diagram: We explored the relationships of the terms in the diagram. We then learned that each of the properties above can be placed into one of two categories: extensive properties or intensive properties. In the previous section, we saw that this type of analysis can be helpful in certain problems.

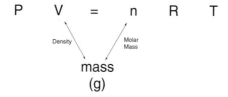

Now we will focus on a different way to categorize the properties above. But this time, one property fits into a category all by itself.

Let's consider which properties above are subject to change. We know that the pressure can change; so can the volume. The number of moles (n) can change in a chemical reaction (and therefore, so can the amount of grams of a specific substance). The temperature can certainly be changed. Density changes with temperature and pressure. BUT, *molar mass* does not ever change. You can change any of the other properties on the diagram above, but *not* the molar mass. The molar mass of CO_2 is ALWAYS 44.01 g/mol. Changing the temperature might affect the density but it will *not* affect the molar mass, because the molar mass is only dependent on the molecular formula, which does not change.

So, we have seen that molar mass is in a category all by itself. Once you appreciate this fact, you will be in a better position to solve a certain type of problem. This problem type is perhaps one of the most difficult types of problems you will ever see. In fact, you may not ever see a problem this difficult, so you can skip this section if it seems too tough. But give it a try, and you might find that you come away with an even deeper understanding of problem-solving methodology.

The type of problem goes like this: You are given a set of conditions *for an unknown gas*. Using the values given, you can calculate the molar mass of the unknown gas. Then, you can use that molar mass in an entirely new set of conditions for a sample of same unknown gas.

Let's see a specific example of this type of problem, and let's work through it slowly.

EXERCISE 4.59.   A 550 mL vessel contains 3.12 g of an unknown gas at a temperature of 25° C and a pressure of 745 torr. What is the density of a small sample of this gas at 50° C and 760 torr?

Answer: This problem is actually giving you two different sets of conditions. The first set of conditions allows you to calculate the molar mass of the gas (the one and only property that is constant no matter what the conditions are). Then, once you have calculated the molar mass, you know more information about the gas. You can think of the two sentences of the problem like this: the first sentence is only there to provide you with information about the gas—specifically, the molar mass. The second sentence is asking you to use the molar mass together with the new conditions to make a calculation.

Don't confuse this with a problem where there is a change of conditions. In the next section of this chapter, we will focus on those types of problems. This problem here does NOT involve a change of conditions. It is not necessarily the same exact molecules that are being heated or subjected to a change in pressure. Rather, there is one set of conditions that tells you something about an unknown gas. Then you are asked to make a calculation regarding that type of gas under different conditions. It is a subtle difference.

Now let's strategize. This problem is solved with two stages. In stage 1, focus on the first sentence of the problem, which gives the following information:

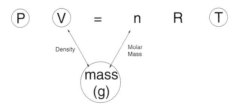

You can use P, V, and T to calculate n. Then you can use n, together with the mass, to calculate the molar mass. That value will be the same under ANY conditions. So, stage 1 of the strategy is to calculate the molar mass

In stage 2 of the strategy, focus on the second sentence of the problem, which gives you a new set of conditions for a sample of the unknown gas. You are only given P and T, but remember that you calculated the molar mass in stage 1 of the strategy. So, you have the following information regarding the second set of conditions:

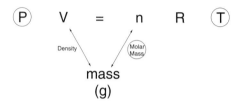

This now resembles the problems from the previous section. You did many problems like this. At first glance, it might seem as though you don't have enough information to solve for anything, and that should trigger your memory about the trick you learned. The values shown above are three of the four intensive properties, and you are asked to calculate the fourth intensive property (density). Therefore, you can choose n = 1, and that will allow you to calculate V and mass. And you can then use those two values to calculate your answer.

Now that you have your two-stage strategy, let's actually solve the problem. In stage 1, begin by doing the appropriate conversions:

$$V = 550 \text{ mL} \times \frac{1 \text{ L}}{1000 \text{ mL}} = 0.550 \text{ L}$$

$$745 \text{ torr} \times \frac{1 \text{ atm}}{760 \text{ torr}} = 0.9803 \text{ atm}$$

(we have included an extra digit that is not a significant digit)

$$T = (25 + 273.15) \text{ K}$$
$$= 298.15 \text{ K}$$

(we have included extra digits that are not significant digits)

Now, plug these values of P, V, and T into the Ideal Gas Law to calculate n:

$$PV = nRT$$
$$n = \frac{PV}{RT}$$
$$= \frac{(0.9803 \text{ atm})(0.550 \text{ L})}{(0.08206 \text{ L atm mol}^{-1} \text{ K}^{-1})(298.15 \text{ K})}$$
$$= 0.02204 \text{ mol}$$

(we have included an extra digit that is not a significant digit, because this is not the final answer)

Then, use that (together with the given mass) to calculate the molar mass:

$$\text{molar mass} = \frac{\text{mass}}{\text{moles}}$$

$$= \frac{3.12 \text{ g}}{0.02204 \text{ mol}}$$

$$= 141.6 \text{ g/mol}$$

(the last digit is not a significant digit, but carry it through until you reach the end of your calculations)

Now stage 1 is complete. You have calculated the molar mass, so you are ready to move on to stage 2. Look at the information in the second sentence of the problem. You are given P and T, which you need to convert before you make any calculations:

$$P = 760 \text{ torr} \times \frac{1 \text{ atm}}{760 \text{ torr}} = 1.00 \text{ atm}$$

$$T = (50 + 273.15) \text{ K}$$
$$= 323.15 \text{ K}$$

When developing this strategy, you chose the value of $n = 1$, and now you can use that to calculate V:

$$PV = nRT$$

$$V = \frac{nRT}{P}$$

$$= \frac{(1 \text{ mol})(0.0821 \text{ L atm mol}^{-1} \text{ K}^{-1})(323.15 \text{ K})}{(1.000 \text{ atm})}$$

$$= 26.53 \text{ L}$$

(once again, the last digit is not a significant digit, but carry it through until you reach the end of your calculations)

Next, calculate the mass. You have n (the value has been set as $n = 1$), and you calculated the molar mass in stage 1. So, you can use the molar mass to convert moles into mass:

$$1 \text{ mol} \times \frac{141.6 \text{ g}}{1 \text{ mol}} = 141.6 \text{ g}$$

You're just about done. You have now calculated V and mass, so you can use those values to calculate the density:

$$\text{Density} = \frac{\text{mass}}{\text{volume}} = \frac{141.6 \text{ g}}{26.53 \text{ L}} = 5.34 \text{ g/L}$$

That was tough stuff because there were so many steps involved. If you had trouble with this, don't get discouraged. This is probably about as tough as it gets. Most of the problems you encounter will be simpler than this.

It might take you some time to really digest this strategy, and to learn how to recognize the problems that require this strategy. But that's expected. Let's quickly review the strategy now.

In these types of problems, the trick is to realize that molar mass is a fixed quantity that does not change under any circumstances. Therefore it can be used as a *bridge* between one set of conditions for an unknown gas and an entirely different set of conditions. In these types of problems, your strategy will have two stages. In stage 1, you will use the first set of conditions to calculate the molar mass. Then, in stage 2, you use the molar mass and the second set of conditions to make your calculations.

Now try to do each of the following problems. You will need to use a blank piece of paper to develop your strategy, do your conversions, and calculate your answer. You will probably need one piece of paper for each problem.

PROBLEM 4.60. A 750 mL vessel contains 2.50 g of an unknown gas at a temperature of 0° C and a pressure of 300 torr. A small sample of this gas is removed from the vessel and placed into a new vessel at 50° C and 760 torr. What is the density of the sample?

PROBLEM 4.61. A 1.00 L container is filled with 2.50 g of an unknown gas at a temperature of 25° C and a pressure of 350 torr. A small sample of gas is removed from the container and is cooled to a temperature of 0° C, and the density is found to be 1.25 g/L. What is the pressure of this small sample of gas?

PROBLEM 4.62. A 750 mL container is filled with 4.5 g of an unknown gas at a temperature of 0° C and a pressure of 760 torr. A small sample of this gas is removed from the container and is subjected to a pressure of 4.50 bar. The density of the gas is measured to be 3.56 g/L. What is the temperature of the gas?

4.6 COMBINED GAS LAW: CALCULATIONS INVOLVING A CHANGE OF CONDITIONS

In this section, we will get practice solving problems that involve *a change of conditions*. Recall that the Ideal Gas Law has four variables: P, V, n, and T. In most problems where a change takes place, you will find that two of the four variables are kept constant, and the other two change.

To see how we deal with changing conditions, we will need to use two equations:

$$P_1V_1 = n_1RT_1$$
$$P_2V_2 = n_2RT_2$$

The first equation describes the first set of conditions, and the second equation describes the second set of conditions. For example, P_1 represents the initial pressure, and P_2 represents the final pressure after the change takes place.

Now we divide one equation by the other, and we get:

$$\frac{P_1V_1}{P_2V_2} = \frac{n_1RT_1}{n_2RT_2}$$

Since R is in the numerator and in the denominator, we can cross off R, and we get:

$$\frac{P_1V_1}{P_2V_2} = \frac{n_1T_1}{n_2T_2}$$

In most problems, you will find that n is one of the variables kept constant. It is possible for the value of n to change (in a chemical reaction, n can change), but most problems do not involve a change of n. Therefore, in most situations $n_1 = n_2$, the above equation can be simplified by crossing off n_1 and n_2, and we get:

$$\frac{P_1V_1}{P_2V_2} = \frac{T_1}{T_2}$$

This equation is called the Combined Gas Law (we got it from combining two gas law equations), and it will be your ticket to solving many problems. You need to know this equation like the back of your hand. But *don't* memorize it. If you try to memorize it, you might forget it. Instead, you should understand how to derive it yourself. We just divided one equation by another, and then we crossed off R and n. Once you understand the process, you should realize that you can always re-derive this equation in a few seconds.

You will find that many equations in your textbook are logical conclusions of the Combined Gas Law. You will learn about Charles' Law, Boyle's Law, and Gay-Lussac's Law. But you don't need to memorize any of these laws. For example, if the temperature is kept constant, then $T_1 = T_2$, so we can cross them off from the above equation, and we get Boyle's Law (it is easy to remember what is held constant in Boyle's Law, because *Boyle's* sounds like *boils* which reminds us that the temperature is held constant in this law):

$$\frac{P_1V_1}{P_2V_2} = \frac{\cancel{T_1}}{\cancel{T_2}}$$

$$\frac{P_1V_1}{P_2V_2} = 1$$

$$P_1V_1 = P_2V_2$$

If instead, we keep the pressure constant, then $P_1 = P_2$. When we cross off P_1 and P_2 from the above equation, we get Charles' Law:

$$\frac{P_1V_1}{P_2V_2} = \frac{T_1}{T_2}$$

$$\frac{V_1}{V_2} = \frac{T_1}{T_2}$$

Finally, if we choose to keep the volume constant, then $V_1 = V_2$, and we get Gay-Lussac's Law:

$$\frac{P_1V_1}{P_2V_2} = \frac{T_1}{T_2}$$

$$\frac{P_1}{P_2} = \frac{T_1}{T_2}$$

The truth is that we have done things somewhat backwards. The textbooks usually teach these three laws first. Then, these laws are combined to derive the Ideal Gas Law (PV = nRT), and finally the Combined Gas Law. It makes sense to teach it in that order, because that is the way the Ideal Gas Law was discovered in the first place (by a combination of the three laws above, which were all discovered through careful experimentation). But in the end, it is often difficult for some students to remember the exact equations for each of these laws. So, instead, we are doing things backwards in this book. If you know the Ideal Gas Law (PV = nRT), which is easy to remember, then you can re-derive everything else, as you did above. You just divide one equation by another, cross off n and R, and end up with the Combined Gas Law. And, you can figure everything else out using the Combined Gas Law.

Now let's see an example of how we can use the Combined Gas Law to solve problems.

EXERCISE 4.63.   A gas has a volume of 2.50 L at a temperature of 40° C. What will the volume of the gas be if its temperature is raised to 75° C while its pressure is kept constant?

Answer: When reading the problem above, you should immediately recognize that a change is taking place (the temperature is raised . . .). This is a signal to use the Combined Gas Law. When you read the problem carefully, you can verify that the number of moles is not changing, so n is constant, and you use the Combined Gas Law:

$$\frac{P_1 V_1}{P_2 V_2} = \frac{T_1}{T_2}$$

Next, look to see if any other terms are constant. You'll see that pressure is kept constant, so you can cross off P_1 and P_2, and you get:

$$\frac{V_1}{V_2} = \frac{T_1}{T_2}$$

This is Charles' law. Now, consider what information you were given. The problem gives you the following values:

$$\frac{\cancel{V_1}}{V_2} = \frac{\cancel{T_1}}{\cancel{T_2}}$$

So, you see that you are given three of the four variables in the equation. You just have to plug and chug to get your answer.

Don't forget to first do the appropriate unit conversions. In this problem, V_1 is given in liters, so you only have to convert T_1 and T_2:

$V_1 = 2.50$ L

$T_1 = (40 + 273.15)$ K $= 313.15$ K

$T_2 = (75 + 273.15)$ K $= 348.15$ K

Before you can just plug and chug, you need to rearrange the equation so that it is in the form of $V_2 = \ldots$.

This requires two algebraic manipulations (which are both simple to do). If you feel comfortable with these manipulations, then you can skip over the next few sentences.

The first manipulation is to multiply the numerator of each side by the denominator of the other side, like this:

$$\frac{V_1}{V_2} \diagdown\!\!\!\!\!\!\diagup\; \frac{T_1}{T_2}$$

That gives you the following equation:

$$V_2 T_1 = V_1 T_2$$

Then, your second manipulation is to divide both sides by T_1 so that you get your equation in terms of V_2:

$$V_2 = \frac{V_1 T_2}{T_1}$$

Now you are ready to plug and chug:

$$
\begin{aligned}
V_2 &= \frac{V_1 T_2}{T_1} \\
&= \frac{(2.50 \text{ L})(348.15 \text{ K})}{(313.15 \text{ K})} \\
&= 2.78 \text{ L}
\end{aligned}
$$

In the example above, two conditions were kept constant (n and P). In some cases, only n is kept constant, which means that P, V, and T are all changing. Let's see an example of this:

EXERCISE 4.64.   A sample of gas at a pressure of 760 torr and in a volume of 2.00 L was heated from 25.0° C to 75.0° C. The volume of the container expanded to 3.56 L. What was the final pressure of the gas?

Answer: Once again, here is a problem where n is not changing (as we have said before, this is very common). But everything else *is* changing. The temperature is raised, the volume expands, and the pressure is also changing. So you will need to use the Combined Gas Law:

$$\frac{P_1 V_1}{P_2 V_2} = \frac{T_1}{T_2}$$

There are six variables in the equation above, and the problems gives you five of them:

$$\frac{\widehat{P_1}\widehat{V_1}}{\widehat{P_2}\widehat{V_2}} = \frac{\widehat{T_1}}{\widehat{T_2}}$$

Start by writing what you are given:

Initial Conditions	Final Conditions
$P_1 = 760$ torr	$P_2 = ?$
$V_1 = 2.00$ L	$V_1 = 3.56$ L
$T_1 = 25.0°$ C	$T_1 = 75.0°$ C

Next, you must do all necessary unit conversions. You get:

Initial Conditions	Final Conditions
$P_1 = \mathbf{1.000\ atm}$	$P_2 = ?$
$V_1 = 2.00$ L	$V_2 = 3.56$ L
$T_1 = \mathbf{298.15\ K}$	$T_2 = \mathbf{248.15\ K}$

Once again, you are using extra digits that are not significant (which you will carry along until you get to your final answer, which should only have three significant digits).

Before you can plug and chug, you need to manipulate the Combined Gas Law, so that it is in terms of P_2. Remember from the last exercise that this requires two manipulations. Manipulation #1 is to multiply the numerator of each side by the denominator of the other side:

$$\frac{P_1 V_1}{P_2 V_2} \times \frac{T_1}{T_2}$$

That gives you the following equation:

$$P_1 V_1 T_2 = P_2 V_2 T_1$$

Then, manipulation #2 is to divide both sides of the equation by V_2 and T_1, which give you the equation in terms of P_2:

$$\frac{P_1 V_1 T_2}{V_2 T_1} = P_2$$

Now you are ready to plug and chug:

$$P_2 = \frac{P_1 V_1 T_2}{V_2 T_1}$$

$$P_2 = \frac{(1.000\ \text{atm})(2.00\ \text{L})(348.15\ \text{K})}{(3.56\ \text{L})(248.15\ \text{K})}$$

$$= 0.788\ \text{atm}$$

Now let's get some practice. Let's review the strategy. When we are dealing with problems where there is a change of conditions, then we will use the Combined Gas Law. In some cases, this equation will be reduced to Charles' Law or Boyle's Law or Gay-Lussac's Law (if one of the conditions, other than n, is kept constant). In other cases P, V, and T will all change. Whatever the case, you should write down all of the information that is given to you in the following format:

Initial Conditions   ***Final Conditions***

$P_1 =$ $P_2 =$

$V_1 =$ $V_2 =$

$T_1 =$ $T_2 =$

Make sure to do any necessary unit conversions, and then you can plug and chug. Compared to the previous section of this chapter, the problems in this section should seem a lot more straightforward (and quicker to solve).

PROBLEM 4.65.   A sample of a gas has a pressure of 750 torr and a temperature of 250° C. To what temperature must the gas be heated in order to double the pressure if there is no change in the volume of the gas?

PROBLEM 4.66.   A sample of a gas has a pressure of 750 torr, a volume of 2.74 L, and a temperature of 0° C. The temperature is raised to 200° C under conditions that let the pressure change to 780 torr. What was the final volume of the gas?

PROBLEM 4.67.   A sample of a gas with a volume of 6.25 L, a pressure of 755 torr, and a temperature of 21° C expanded to a volume of 9.82 L and a pressure of 423 torr. What was its final temperature?

4.7 IDEAL GAS LAW AND STOICHIOMETRY PROBLEMS

In Chapter 3, we learned how to use a balanced reaction to relate the mass of the reactants to the mass of the products. In every case, we said that the trick was to convert everything into moles. A typical problem from that chapter looked something like this:

Consider the following reaction:

$$C_3H_8 \ (g) + 5 \ O_2 \ (g) \longrightarrow 3 \ CO_2 \ (g) + 4 \ H_2O \ (g)$$

How many ***grams*** of CO_2 can be made from 100.0 g of C_3H_8?

In this problem, you would begin by converting *grams* of C_3H_8 into *moles* (using the molar mass of C_3H_8 as a conversion factor). Once you know how many moles of C_3H_8 you have to work with, then you can use the balanced reaction to calculate how many moles of CO_2 will form. And finally, you would use the molar mass of CO_2 to convert *moles* of CO_2 back into *grams*. This strategy can be visually represented using the following diagram:

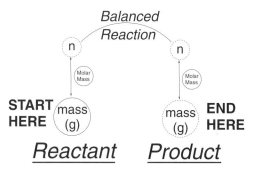

That was the basic idea. We saw many variations on the same theme, but in every case, ***the key to solving those problems was converting mass into moles and then using the balanced reaction to determine the ratio of moles of reactants to moles of products***.

Now, with the tools that we have learned in this chapter, we can appreciate that there is a larger picture here. Using the diagrams that we have used throughout this chapter, we can see that the problems in Chapter 3 followed a very narrow path on a larger map of possible routes:

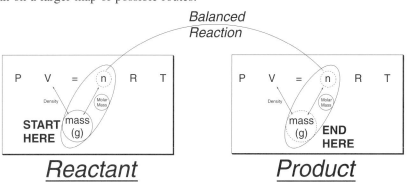

We are now using two diagrams connected by a *bridge*. The first diagram shows the properties of a *reactant* in a particular reaction. The second diagram shows the properties of a *product* of the reaction. Notice that there is a *bridge* between these two diagrams connecting the values of *n* in each diagram. That bridge exists because of the balanced reaction. With the balanced reaction, you can determine how many moles of product are produced from every one mole of reactant. So, with a balanced reaction, you can connect the diagram of a reactant with the diagram of a product.

And, now, we can better appreciate what we were doing in Chapter 3 on stoichiometry. Using a real-life analogy, imagine that you are familiar with a route to get from one place in a city to another (go two blocks, turn right, then go one block, turn right, etc.), but you are *not* familiar with the neighborhood at all. If someone were to plop you down just one block off of the path, you would have no idea where you were. Then imagine that, over time, you become more familiar with the neighborhood. At that point, you realize that there are multiple paths that will take you to the same place. Most of us have experienced this kind of thing at some point in our lives. So, we can use our life experience to relate to what is going on right now in this chapter.

Now that we have become more familiar with the neighborhood, look at the double-diagram above and you will see (in the ovals) the route we traveled when we were doing stoichiometry problems in Chapter 3. At the time, it was the only route we knew. But now, we can appreciate that there are other ways to dress up a stoichiometry problem. For example, consider the following scenario. As usual, we will use circles to indicate which information is given:

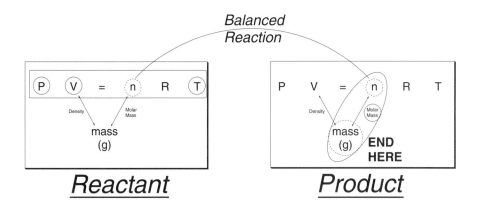

In the above scenario, we are given the following information:

- P, V, and T of the reactant,
- the balanced reaction, which acts as our bridge, and
- the molar mass of the product (which we calculate from its molecular formula).

The problem asks us to calculate the *mass* of the product. So, we would begin by calculating n of the reactant using the Ideal Gas Law. Then, we would use the balanced reaction to determine how many moles of product are formed. And finally, we would use the molar mass of the product to convert moles of product into grams of product.

So you can think of the bridge as a bottleneck. No matter what path you take in a stoichiometry problem (whether you convert into **moles** using molar mass, or whether you calculate **moles** using P, V, and T), you will always cross the bridge, which connects *n* of one compound to *n* of another compound.

Let's consider one more example, and make sure that we can understand the logic. Imagine that we are asked to calculate how many grams of a *reactant* are necessary to produce a given amount of liters of *product*. We are given the following information:

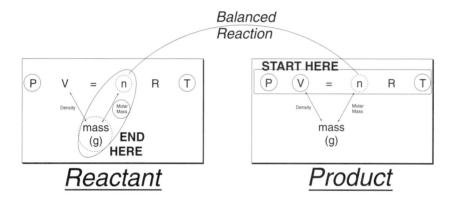

First, we need to calculate the moles of product, using the Ideal Gas Law on the left side of the diagram above. Next, we would use the balanced reaction to determine how many moles of reactant were necessary to make that many moles of product. And finally, we use the molar mass of the reactant to calculate the mass of reactant necessary.

Now let's see how this can be applied to a real problem.

EXERCISE 4.68.   Consider the reaction between H_2 and O_2 shown below:

$$2\ H_2 + O_2 \longrightarrow 2\ H_2O$$

A 15.0 L balloon contains H_2 at a pressure of 740 torr and a temperature of 25° C. The balloon is ignited in the presence of excess oxygen gas. If we assume that all of the hydrogen gas is consumed and converted into water, how many grams of water will be produced?

Answer: Remember that in any stoichiometry problem, you want to deal with everything in terms of moles. That is your bridge. Draw a diagram to see what you have:

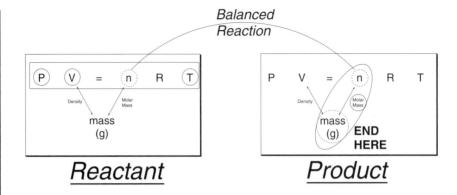

Once you draw everything in one large diagram, it becomes much easier to "see" the strategy that you will need to use. The problem gives you P, V, and T for the hydrogen gas, so you can use the Ideal Gas Law to calculate n (on the left side of the diagram above). Then, use the balanced reaction to calculate how many moles of product will be formed (follow the bridge over to the right side of the diagram above). Finally, use the molar mass of the product to convert moles of product into grams.

Now you have your strategy. But don't forget to do the necessary unit conversions first:

$$740 \text{ torr} \times \frac{1 \text{ atm}}{760 \text{ torr}} = 0.9737 \text{ atm}$$

$$T = (25 + 273.15) \text{ K}$$
$$= 298.15 \text{ K}$$

Now, you are ready to start making calculations. Your strategy is to first use the Ideal Gas Law to calculate moles of reactant:

$$PV = nRT$$

$$n = \frac{PV}{RT}$$

$$= \frac{(0.9737 \text{ atm})(15.0 \text{ L})}{(0.08206 \text{ L atm mol}^{-1} \text{ K}^{-1})(298.15 \text{ K})}$$

$$= 0.5970 \text{ mol}$$

(once again, the last digit is not a significant digit, but carry it through until you reach the end of your calculations)

Now you know how many moles of H_2 are in the balloon. Next, use the balanced reaction as a bridge to get you from moles of reactant to moles of product. The balanced reaction tells you that 2 moles of hydrogen produce 2 moles of water. So, if you start with 0.597 moles of reactant, then you will form 0.597 moles of product

The last step is to convert 0.597 moles of water into grams. To do this you need the molar mass of water:

$$\text{Molar mass of } H_2O = 2(1.008 \text{ g/mol}) + (16.00 \text{ g/mol})$$
$$= 18.02 \text{ g/mol}$$

Using the molar mass as a conversion factor, calculate your answer:

$$0.5970 \text{ mol } H_2O \times \frac{18.02 \text{ g } H_2O}{1 \text{ mol } H_2O} = \underline{10.8 \text{ g } H_2O}$$

(notice that the final answer has three significant figures).

Now, let's get some practice by doing some problems. In some cases, you might be able to "see" the strategy without even drawing the diagram. If so, then great! But, if you are having trouble, you can always use a separate piece of paper to draw a diagram and circle what you are given to help you strategize.

PROBLEM 4.69.   A canister contains 25.5 L of propane at a pressure of 760 torr and a temperature of 25° C. All of the contents of the canister are allowed to react with excess oxygen to produce carbon dioxide and water:

$$C_3H_8 \ (g) + 5 \ O_2 \ (g) \longrightarrow 3 \ CO_2 \ (g) + 4 \ H_2O \ (g)$$

Assuming a complete combustion process, how many grams of CO_2 are produced?

PROBLEM 4.70.   A canister contains butane at a pressure of 760 torr and a temperature of 25° C. All of the contents of the canister are allowed to react with excess oxygen to produce carbon dioxide and water:

$$2 \ C_4H_{10} \ (g) + 13 \ O_2 \ (g) \longrightarrow 8 \ CO_2 \ (g) + 10 \ H_2O \ (g)$$

The butane is completely consumed in the reaction, and 46.5 grams of CO_2 are produced. What is the volume (in liters) of the canister that contained the butane?

PROBLEM 4.71.   Under conditions of high pressure, acetylene (C_2H_2) can react with hydrogen to produce ethane:

$$C_2H_2 \ (g) + 2 \ H_2 \ (g) \longrightarrow C_2H_6 \ (g)$$

How many liters of hydrogen, at 760 torr and 25° C, are needed to react with 15.6 g of acetylene?

(Hint: In this problem, your diagram will not have a bridge that connects a *reactant* with a *product*. Instead, your diagram should have a bridge that connects moles of *one reactant* with moles of *the other reactant*. The balanced reaction above tells you that the mole ratio of the reactants is 1:2. And that information serves as your bridge.)

Answer: _____

4.8 IDEAL GAS LAW AND LIMITING REAGENTS

In the previous section, we saw that the trick for solving stoichiometry problems is to "see" the bridge that connects the number of moles of reactants and products. Any stoichiometry problem will involve crossing this bridge. That means that, at some point during your strategy for solving the problem, you must convert every-thing into moles (n).

In Chapter 3, we saw how to calculate n when we are given the mass (using the molar mass as a conversion factor):

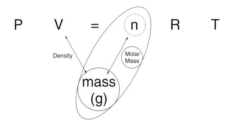

In this chapter, we expanded our horizons a bit, and we saw that we can also calculate n using the Ideal Gas Law (if we have the values for P, V, and T):

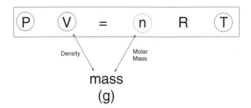

So, we have now seen the two most common ways to calculate n when do-ing a stoichiometry problem. If you go back to the double diagrams in the pre-vious section (the diagrams showing the bridge between a reactant and a prod-uct), you will see that every calculation involving n used one of these two approaches. Every calculation was either done via the molar mass, or via the Ideal Gas Law.

It is possible to calculate n in more convoluted ways, but these ways are less common in typical problems. As an example, imagine that you are given the following three properties for a gas: V, density, and molar mass. We can use the diagram to clearly see the strategy we would use if we wanted to cal-culate n:

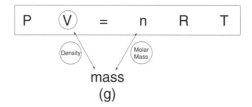

We would first use the volume and density to calculate the mass. Then we would use the mass to calculate n (using molar mass as a conversion factor).

There are other scenarios that we could also imagine (for calculating n), but we will focus on the two most common approaches: using molar mass or using the Ideal Gas Law:

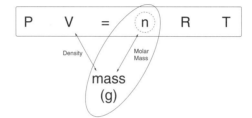

We will use these two approaches to solve limiting reagent problems. These problems are exactly the same as the problems we did toward the end of Chapter 3. So, if you don't remember those problems, you should definitely go back and review them now. The problems you will do here will just have one additional aspect. In order to convert everything into moles (to cross the bridge), you will now have *two* ways to do it: using molar mass *or* using the Ideal Gas Law. In Chapter 3, we only used the molar mass approach. Now, we will mix and match. In fact, you might see a problem where you have to use both approaches.

As an example, suppose two compounds (A and B) react together to form some products. In Chapter 3, a typical problem would have given you the grams of A, the grams of B, and a balanced reaction. Then you would have had to convert everything into moles (using the molar masses of A and B). But now, in this chapter, you might see a problem where you are given the mass of A; and instead of being given the mass of B, you are given the pressure, volume, and temperature of B. So, you will calculate moles of A using the molar mass of A, and you will calculate moles of B using the Ideal Gas Law.

For these problems, you will not use diagrams showing the bridges (like you did in the previous section). Although these diagrams could be helpful, sometimes you can have a problem where you need to consider *three* different compounds (perhaps you are given information about two reactants and one product, like in the following example). A diagram showing all three compounds will be too much information to comfortably fit into one diagram. So, instead, use the diagrams that focus on just one compound at a time. And, you should realize that these diagrams are linked.

EXERCISE 4.72. Consider the combustion of methane, shown below:

$$CH_4\ (g) + \mathbf{2}\ O_2\ (g) \longrightarrow CO_2\ (g) + \mathbf{2}\ H_2O\ (g)$$

Calculate the maximum number of milliliters of CO_2, at 750 torr and 22° C that can be formed from the combustion of methane (CH_4) if 450 mL of methane at 705 torr and 20° C are mixed with 725 mL of O_2 at 680 torr and 80° C.

Answer: In order to calculate the maximum amount of product that can be formed, you will need to calculate which reactant is the limiting reagent (methane or oxygen). Recall from Chapter 3 that the limiting reagent determines how much product is formed. So, you need to calculate the number of moles of methane and oxygen that you have.

Start with methane. You are given P, V, and T for methane:

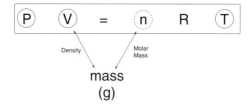

Use the Ideal Gas Law to calculate n. Don't forget to do your conversions first:

$$V = 450\ \text{mL} \times \frac{1\ L}{1000\ \text{mL}} = 0.450\ L$$

$$P = 705\ \text{torr} \times \frac{1\ \text{atm}}{760\ \text{torr}} = 0.9276\ \text{atm}$$

(once again, the last digit is not a significant digit, but carry it through until you reach the end of your calculations)

$$T = (20 + 273.15)\ K$$
$$= 293.15\ K$$

(the last two digits are not significant digits, but carry them through until you reach the end of your calculations)

Now, plug these values of P, V, and T into the Ideal Gas Law to calculate the number of moles of methane:

$$PV = nRT$$
$$n = \frac{PV}{RT}$$

$$= \frac{(0.9276 \text{ atm})(0.450 \text{ L})}{(0.08206 \text{ L atm mol}^{-1} \text{ K}^{-1})(293.15 \text{ K})}$$
$$= 0.01735 \text{ mol}$$

(once again, the last digit is not a significant digit, but carry it through until you reach the end of your calculations)

Next, do the exact same analysis for oxygen. You are given P, V, and T for oxygen, so again, you can use the Ideal Gas Law to calculate n. Don't forget to do your conversions:

$$V = 725 \text{ mL} \times \frac{1 \text{ L}}{1000 \text{ mL}} = 0.725 \text{ L}$$

$$P = 680 \text{ torr} \times \frac{1 \text{ atm}}{760 \text{ torr}} = 0.8947 \text{ atm}$$

(once again, the last digit is not a significant digit, but carry it through until you reach the end of your calculations)

$$T = (80 + 273.15) \text{ K}$$
$$= 353.15 \text{ K}$$

(the last two digits are not significant digits, but carry them through until you reach the end of your calculations)

Now, plug these values of P, V, and T into the Ideal Gas Law to calculate the number of moles of oxygen:

$$PV = nRT$$

$$n = \frac{PV}{RT}$$
$$= \frac{(0.8947 \text{ atm})(0.725 \text{ L})}{(0.08206 \text{ L atm mol}^{-1} \text{ K}^{-1})(353.15 \text{ K})}$$

(once again, the last digit is not a significant digit, but carry it through until you reach the end of your calculations)

Now, you can compare the number of moles of each reactant. You calculated that you have 0.01735 moles of methane and 0.02238 moles of oxygen. On the surface, you would say that you have less methane, so methane should be the limiting reagent. But remember how this was done in Chapter 3. You have to look at the balanced reaction.

The balanced reaction tells you that 1 mole of methane will produce 1 mole of CO_2. However, 1 mole of oxygen will only form half of a mole of CO_2 (look carefully at the balanced reaction). Therefore, you have enough methane to potentially make **0.01735 mol CO_2**. And you have enough O_2 to potentially make $(^1/_2)(0.02238)$ mol) = **0.01119 mol CO_2**. Therefore, O_2 is the limiting reagent.

The limiting reagent determines how much product you get. So, you can only make 0.01119 mol CO_2.

If the problem had asked for a calculation of the maximum number of *moles* of CO_2 that could have been produced, then you would have been finished with the problem. But the problem asked you to calculate the maximum number of *milliliters* of CO_2 that can be formed if the CO_2 is at 750 torr and 22° C. To see how to do this, draw a diagram that shows you what you have so that you can strategize. For CO_2, you have:

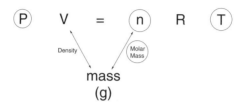

The molar mass was circled, because you can easily calculate it from the molecular formula CO_2. But when you inspect the diagram above, you realize that you will not need to use the molar mass of CO_2. You want to calculate V, and you have P, n, and T. So just use the Ideal Gas Law. Once again, you always have to do the necessary unit conversions before any calculation:

$$P = 750 \text{ torr} \times \frac{1 \text{ atm}}{760 \text{ torr}} = 0.9868 \text{ atm}$$

(the last digit is not a significant digit, but carry it through until you reach the end of your calculations)

$$T = (22 + 273.15) \text{ K}$$
$$= 295.15 \text{ K}$$

Now, just plug and chug, solving for V:

$$PV = nRT$$
$$V = \frac{nRT}{P}$$
$$= \frac{(0.01119 \text{ mol})(0.08206 \text{ L atm mol}^{-1} \text{ K}^{-1})(295.15 \text{ K})}{(0.9868 \text{ atm})}$$
$$= 0.275 \text{ L}$$

(now that you are done with your calculations, you express your answer with the correct number of significant digits).

One last subtle point: the problem asked for the answer in mL, not L. So you need to convert your answer from L to mL:

$$V = 0.275 \text{ L} \times \frac{1000 \text{ mL}}{1 \text{ L}} = 275 \text{ mL}$$

Now, try to get some practice. For each problem, you will need to use a blank piece of paper to develop your strategy, do your conversions, and make all of your calculations. You will probably need one piece of paper for each problem.

PROBLEM 4.73.   Consider the combustion of ethane shown below:

$$2 \ C_2H_6 \ (g) + 7 \ O_2 \ (g) \longrightarrow 4 \ CO_2 \ (g) + 6 \ H_2O \ (g)$$

How many milliliters of CO_2 (measured at 750 torr and 22° C) can be formed if 32.05 g of ethane are mixed with 155 L of O_2 (measured at 760 torr and 25° C)?

PROBLEM 4.74.   Consider the reaction below that shows the formation of ammonia from hydrogen and nitrogen:

$$N_2 \ (g) + 3 \ H_2 \ (g) \longrightarrow 2 \ NH_3 \ (g)$$

How many grams of NH_3 can be formed when 25.0 L of N_2 (measured at 760 torr and 25° C) react with 58.5 L of H_2 (measured at 850 torr and 20° C)?

PROBLEM 4.75.   Consider the reaction below that shows the formation of water from hydrogen and oxygen:

$$2 \ H_2 \ (g) + O_2 \ (g) \longrightarrow 2 \ H_2O \ (g)$$

How many liters of H_2O (measured at 740 torr and 100° C) can be formed when 18.5 L of H_2 (measured at 740 torr and 25° C) react with 30.4 g of O_2?

PROBLEM 4.76.   Consider the following **unbalanced** reaction showing the combustion of hexane:

$$C_6H_{14}\ (l) + O_2\ (g) \longrightarrow CO_2\ (g) + H_2O\ (g)$$

How many grams of CO_2 can be formed when 75.0 g of hexane react with 200.0 L of O_2 (measured at 750 torr and 20° C)?

CHAPTER **5**

ENERGY AND ENTHALPY

This chapter is the first part of a much larger topic called *thermodynamics*. You will revisit thermodynamics in more detail during the second semester of chemistry. The topics in this chapter will lay the foundation that you need for the second semester, so it is important to master the terms, concepts, and problem-solving techniques in this chapter. If you don't get these topics down now, you will find yourself struggling with thermodynamics next semester.

In one sentence, thermodynamics is the study of energy and its interconversions. Put more simply, thermodynamics is the study of *how*, *why*, and *when* energy can be transferred from one place to another. In this chapter we focus on *how* energy is transferred. In the second semester of chemistry, you will learn about *why* and *when* energy is transferred (entropy and free energy).

The first half of this chapter will focus on theory, terminology, and analogies. The second half of the chapter will focus on problem-solving techniques.

5.1 ENERGY

We will start off our discussion with the different types of energy, but as we do so, keep in mind that we have still not defined what energy really is. We will get to the definition a bit later.

Energy can be classified into the following categories: kinetic energy and potential energy. *Kinetic energy* is energy associated with motion (or velocity), and *potential energy* is energy associated with position.

Let's start with kinetic energy. When a soccer ball is in motion, it has kinetic energy (you might even remember the term $\frac{1}{2} mv^2$ from your high school physics class). When the soccer ball hits another ball, it will transfer some of its energy to the other ball. Molecules can do the same thing. A molecule in motion has kinetic energy that it can transfer when it collides with another molecule. In this chapter, we will spend most of our time focusing on a new concept, called *heat of reaction*, which we will see is a measure of the transfer of *kinetic energy*.

Potential energy is the energy associated with position. For example, if you lift an object high in the air, then it has potential energy associated with its height in the gravitational field of the earth. If you let go of the object, it will fall down and smash on the ground (the potential energy is being released, and converted into kinetic energy). There are many ways to have potential energy. We just saw that an object in a gravitational field has potential energy. Two charged particles separated by a distance also have potential energy. This is the kind of potential energy that you will learn more about in the second semester (electrochemical potential energy). At this point in the game, we will focus our discussions on kinetic energy.

When we talk about the total energy of a system, we are referring to the *sum total* of all kinds of energy (the sum of the kinetic energy *and* every kind of potential energy) for each and every particle in the system. We use the symbol **E** to refer to this sum total of all of the energy in a system (some textbooks use the symbol **U**).

The energy of a system can only change when the system exchanges energy with the surroundings. So, if we are talking about a chemical reaction, then the reaction flask is the system, and the surroundings are the rest of the universe. However, practically speaking, we often refer to the surroundings as the immediate environment affected by the system.

When a system exchanges energy with its surroundings, there is a change in the energy of the system. To show that a change took place, we use the symbol Δ (delta). So, ΔE (pronounced: delta E) refers to the *change* in energy of the system. To see how the system and its surroundings are affected when they exchange energy, let's consider an analogy.

Suppose you are giving money to a friend. If you give your friend 10 dollars, then you have lost 10 dollars and your friend has gained 10 dollars. In other words, your friend gained *exactly* the amount that you lost. This is true even if you do more than one transaction. For example, imagine that you give your friend 10 dollars, and then your friend gives you back 2 dollars. At the end of the day, you have lost 8 dollars, which means that your friend must have gained 8 dollars. Once again, your friend gained *exactly* the amount that you lost. This makes sense, because the total amount of money (the amount that you had originally plus the amount that your friend had originally) will always remain the same. In other words, the total amount of money is conserved.

This simple concept applies to the exchange of energy between a system and its surroundings, and we call this the *First Law of Thermodynamics*. This law tells us that energy is conserved. Energy is never created or destroyed; rather, it is just transferred from one place to another. Therefore, the amount of energy gained by the surroundings will be *exactly* the same as the amount of energy lost by the system. This simple idea can be summarized in the following equation:

$$\Delta E_{surroundings} = -\Delta E_{system}$$

Notice the negative sign on one side of the equation. We put in this negative sign because ΔE not only measures the change in energy, but it also measures the *direction* of the change. Let's go back to our money analogy. If you give your friend 8 dollars, then your change will be -8 dollars, and your friend's change will be $+8$ dollars. The amount is exactly the same, but the sign is different (one is a positive change and the other is a negative change). So, for any physical process, you will have two values ($\Delta E_{surroundings}$ and ΔE_{system}); one of these values will be a positive number and the other will be a negative number. To show that these values have opposite signs, we place a negative sign in our equation.

If we are going to talk about a system losing or gaining energy (at the expense of the surroundings), then we will need to know the units we use to measure energy. We can't just say that the system lost 5. Then we would have to ask: 5 what? Remember what we said in Chapter 1. The units are just as important as the number before the units. We can easily determine the units of energy by looking at an expression that we have seen before; kinetic energy is $\frac{1}{2}$ mv^2. The units for this expression are:

$$\text{mass} \times \frac{\text{distance}}{\text{time}} \times \frac{\text{distance}}{\text{time}}$$

Let's remember what the SI units are for mass, distance, and time. We saw that the SI unit for mass is a kilogram (kg), the SI unit for distance is a meter (m), and the SI unit for time is a second (s). Therefore, the units of energy must be:

$$\text{kg} \times \frac{\text{m}}{\text{s}} \times \frac{\text{m}}{\text{s}}$$

Using exponents, we can rewrite the units above like this:

$$\text{kg} \times \text{m}^2 \times \text{s}^{-2}$$

These are the units of energy. To save time, so that we don't have to write out the expression above every time we express an amount of energy, we define a new term, called a **Joule**:

$$1 \text{ Joule} = 1 \text{ kg} \times \text{m}^2 \times \text{s}^{-2}$$

So from now on, we will measure energy in units of Joules (J) and kiloJoules (kJ), where

$$1 \text{ kJ} = 1000 \text{ J}$$

There is one other type of unit that you should be familiar with. It is an older unit for describing energy, but it is still in use, so you will need to be familiar with it. A calorie is also a measurement of energy, and it is based on how much energy it takes to raise 1 gram of water by 1° C. In general, you will work with Joules, rather than calories, and you will express your answers using units of Joules. So, you will need to know how to convert between Joules and calories. The conversion factor is:

$$1 \text{ calorie} = 4.184 \text{ J}$$

With this conversion factor, we can use the factor-label method to convert calories into Joules and vice versa. Let's see an example:

EXERCISE 5.1.   Convert 152.4 calories into units of kJ.

Answer: You will need to convert calories into Joules, and then Joules into kiloJoules. By using the factor-label method, you can do everything in one step:

$$152.4 \text{ cal} \times \frac{4.184 \text{ J}}{1 \text{ cal}} \times \frac{1 \text{ kJ}}{1000 \text{ J}} = 0.6376 \text{ kJ}$$

Notice that your answer has four significant figures.

These conversions are not difficult. In fact, they require the same skills used in Chapter 1, when you were practicing the factor-label method. You need to know how to do these types of conversions, because many problems will require you to convert from calories to Joules (or vice versa) as a first or final step in solving the problem. These problems will just assume that you can do this kind of conversion. So, let's get a little bit of practice. In the next section, we will do some problems where the first or last step will be to do a conversion like this.

For each of the conversions below, use the factor-label method to determine your answer. Don't forget to count significant figures before recording your answer.

PROBLEM 5.2.   37.2 J = _____ cal

PROBLEM 5.3.   104.6 kJ = _____ kcal

PROBLEM 5.4.   0.46 kJ = _____ cal

PROBLEM 5.5.   637.2 J = _____ kcal

PROBLEM 5.6.   29.4 cal = _____ J

PROBLEM 5.7.   108.2 kcal = _____ kJ

PROBLEM 5.8.   0.043 kcal = _____ J

PROBLEM 5.9.   892.1 cal = _____ kJ

So far, we have categorized the different types of energy (kinetic and potential energy), and we talked about the units of energy. But we still have not defined what energy is. The best definition for energy goes like this: *energy is the ability to do work or produce heat*. This might seem like a nebulous definition. But, in the next section we will learn about the ways that energy can be transferred, and we will see that this definition actually makes a lot of sense.

5.2 ENERGY TRANSFER: WORK AND HEAT

In the previous section, we saw an analogy of two friends giving each other money, and we saw that the total amount of money remained constant. If two friends walk into a room and each of them is carrying 100 dollars, then the total amount of money that they have is 200 dollars. They can give each other money back and forth as many times as they like, but there will always be a total of 200 dollars between them. This is true whether they give each other money in one transaction or in many transactions.

We said that the same is true when a system exchanges energy with its surroundings. We saw that energy is conserved. You might imagine that the calculations could get tedious if a system was able to exchange energy with its surroundings in 100 different ways. You would have to add up all of the transactions to know how the system was affected. But, there is some very good news. It turns out that there are only *two* ways to transfer energy. That's right, just two: work and heat. So let's look at each of them in detail, and then we can talk about the interplay between them.

Heat is the term used to describe the following situation: when a system and its surroundings have different temperatures, energy is transferred from the area of high-temperature to the area of low-temperature. This transfer of energy is called *heat*. The terms *heat* and *temperature* are often confused, so it is worth taking a few minutes here to explain the difference between them.

To help see the difference between heat and temperature, we will use another money analogy. Imagine that you have a rich uncle who sends you money via a wire transfer every month. At any moment in time, you can ask how many dollars you have in your account. But it does not make sense to ask how many wire transfers are *in* your account. A wire transfer is not something that can exist in your account—it is a term that describes the transfer of money from one account to another account. Heat is like a wire transfer. You can't ask how much heat is in a system—that doesn't make any sense. You can only ask how much energy the system has. If you ask specifically how much kinetic energy the system has, then you are talking about temperature.

Our analogy helps explain that heat is a measure of the ***transfer*** of kinetic energy from one system to another. If you have only one system that is isolated from the rest of the universe, then it makes no sense to talk about heat. Heat does not apply to only one system. It only applies when two systems come into contact with each other, and energy is transferred from one to the other. For example, when you put a hot bowl of soup into a refrigerator, the soup molecules transfer energy to the air molecules—the soup cools down and the cold air warms up a bit.

There are many reasons why people confuse heat and temperature. For example, we just talked about a "hot" bowl of soup. This means the object has a high temperature; it does ***not*** mean that the object has a lot of heat.

Before we move on to the second method of energy transfer (work), we need to see how we represent heat in equations. We use the letter "q" to refer to heat. We will see "q" in equations very soon. When you have a system in contact with its surroundings, the sign of q (+ or −) tells you whether energy flows from the system to the surroundings or vice versa. Because chemists care about the system and not so much about the surroundings, we have adopted the convention that the sign of q indicates what happens *to the system*. If q is negative, that means the system loses energy. Because energy flows out of the system, we call the process exothermic. On the other hand, if q is positive, that means the system gains energy. Because energy flows into the system, we call the process endothermic. So, the sign of q tells you whether a reaction is endothermic or exothermic.

The convention that chemists use is different than the convention used by engineers. Engineers build objects that are supposed to provide energy *to the surroundings*. Since engineers care more about the surroundings than the system, they have adopted the opposite convention regarding the sign of q. For engineers, a positive value of q means *the surroundings* are gaining energy, rather than the system.

We have argued before that there are only two ways to transfer energy: heat and work. We have talked about heat, and now we need to talk about work.

Before we see how chemical reactions can do work, we need to define what work is in general. You know from experience that pushing a heavy table across a floor requires you to do some work. If we had to quantify how much work you need to do, we would say that it depends on two things: how hard you have to push, and how far you need to push. We can express work like this:

(The work required to push the table) =
(the amount of force it takes to push the table) ×
(the distance you want to push the table)

Notice how many words it took to say such a simple concept. That's why we use equations. We can express the relationship more simply and quickly with an equation:

$$w = F \times \Delta d$$

The Δ tells us that we only care about the change in distance (how far you actually had to push). Imagine that you were pushing a table along a highway, and I told you to push the table to highway marker #321. If your starting point is highway marker #320, then you only have to push for 1 mile, not 321 miles. All that matters is how far you have to push (Δd).

The equation above is a simple idea, but very powerful. It should ***not*** be memorized. It should make sense. This particular equation probably already made sense to you, but I point this out now, because other equations (that you are about to see) might not make so much sense unless you make an effort to understand them. As soon as you resort to rote memorization, you will be in trouble.

Now that we have a general definition for work, we can ask what *work* means for a process, such as heating a gas. To see this better, imagine a canister with a movable piston:

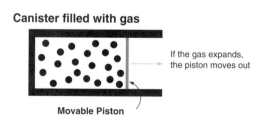

Canister filled with gas

If the gas expands, the piston moves out

Movable Piston

As we heat the gas inside the canister, the gas expands, so the piston moves (pushing on the atmosphere). Remember the equation we just saw: work = (force) × (change in distance). So, we need to know the force and the distance in order to calculate the work done. Distance is easy to understand (how many meters does the piston move), so let's focus on the force required. If there was *no* atmosphere applying pressure on the piston, then it would take no force at all to push the piston (ignoring friction). But in the presence of atmospheric pressure, a force will be required to push the piston against this external pressure. For a gas, the force required would be the atmospheric pressure multiplied by the area of the piston (this should make sense because a wider piston will have more surface area, and therefore the atmosphere will be pushing on it more). So, the relationship is:

(The force required for the expanding gas to push against the atmosphere) = (atmospheric pressure) × (the surface area of the piston)

Again, we will put this relationship into a simple equation:

$$F = P \times A$$

Now that we have an expression for the force required to push against the external pressure, we can use the previous equation to calculate the *work* required:

$$w = F \times \Delta d$$

And if we replace F with (P × A), then we get:

$$w = P \times A \times \Delta d$$

We know that area multiplied by distance is just volume, so we can rewrite the equation as:

$$w = P \times \Delta V$$

We are almost done. We just need to add a negative sign to the equation. To see why, let's focus on ΔV. ΔV can be negative or positive. If the volume increases, then ΔV will be positive. If the volume decreases, then ΔV will be negative. So, we can see that w can be negative or positive. But think about what w means. Remember that we have chosen the convention of always focusing on the system (unlike engineers who focus on the surroundings). When ΔV is positive, that means that the gas is expanding, and the gas is doing work on the surroundings. So, the system is transferring energy to the surroundings, which means the system is losing energy. But when the system loses energy, w should be negative (because that is the convention we chose). So according to our convention, a positive ΔV is supposed to produce a negative w. And we show that by putting a negative sign in the equation:

$$w = -P \times \Delta V$$

And now we have a definition for work. It is simple, but VERY important. In the rest of this chapter, we will be talking about the $P\Delta V$ term a lot. So you need to understand why this term represents a transfer of energy in the form of work.

So far, we have seen how an expanding gas can do work (by pushing on the atmosphere), and now we can begin to understand how a chemical reaction can do work. When chemicals react, they can often produce products that are gases. These gases push against the entire atmosphere, which has to make room for the new gases that have been formed. So the gases are doing work by pushing on the atmosphere. This is true even if the reaction is not taking place in a canister with a piston. The canister and piston were just helpful teaching tools so that we could see what was happening. If you mix baking soda and vinegar, they react to produce carbon dioxide, which is a gas (this is a common experiment that most of us have done at some point with household items). The carbon dioxide that is formed is pushing on the entire atmosphere as it is being formed. It is not too much different than trying to blow up a balloon underwater. It is hard to do because you are fighting against the pressure of the water. The atmosphere does not provide as much pressure as the water in the pool, but the concept is similar.

To review, we have seen so far that there are only two ways to transfer energy: heat and work. We have seen both of these, so now we are ready to make a new statement (a new equation). If the energy of a system is changing, then it must be that the system transferred energy with its surroundings either in the form of work, or heat, or a combination of both. This concept can be summarized in a simple equation:

$$\Delta E = q + w$$

This equation is just restating what we have seen so far. In order for a system to change its amount of energy, it must exchange energy with the surroundings, and there are only two ways to do that: heat and work.

Before we can move on, we should point out the difference between work and heat. Both are ways of transferring energy, but one is microscopic and the other is macroscopic. Think about what is happening when energy is transferred in the form of heat. Let's go back to the example of the hot bowl of soup in the refrigerator. At the surface, the soup molecules are moving very fast and are colliding with the slow-moving air molecules. The collisions transfer energy from the soup molecules to the air molecules. The air molecules pick up speed, and the soup molecules slow down a bit. This continues to happen until the soup is the same temperature as the surrounding air. During this process, each molecule of soup is traveling in its own direction with its own speed. If you were the size of a molecule, you would see a lot of motion. But, on a macroscopic level, there is no motion at all. You don't see the soup moving. It seems to be still, even though each molecule is moving quite fast.

Now consider a river. The water is flowing. You put a stick in the river, and the flowing water pushes the stick. Work is being done on the stick. Think about what is happening on a molecular level. All of the water molecules in the river are moving uniformly in one direction and pounding against the stick. This "collective" pounding is responsible for the force that is doing the work of pushing the stick. On a macroscopic level, you can actually see something happening: the river is flowing, and the stick is being pushed.

Now we can appreciate the difference between work and heat. Heat is energy being transferred on a microscopic level, and work is energy being transferred on a macroscopic level. And those are the only two ways to transfer energy.

Before we see some problems, let's quickly revisit the definition that we gave earlier for energy. We saw that energy is *the ability to do* **work** *or produce* **heat**. That definition might have seemed a bit nebulous before. But now, with everything we have just discussed about the only two ways there are to transfer energy, doesn't that definition make a bit more sense? We can now appreciate that the definition seems to be quite appropriate. BUT, let me leave you with the following parting thought.

Imagine that I tell you that I have a lot of bafoofs, and you ask me: "What are bafoofs?" I answer your question by saying: "I can kick them at you or I can throw them at you. Which way would you like them?" Have I answered your question? No—I have not. Go back to our definition of energy. Work and heat are the two ways to *transfer* energy, but we still haven't said what energy *is*. When we say that energy is the ability to do work or produce heat, we are really saying that energy is the ability to transfer energy (either microscopically or macroscopically). In fact, coming up with a definition for energy itself is VERY challenging. Think about that for a bit, but not too much, because we have a lot more to cover, and this is not a book on philosophy. . . .

Now, let's do some simple problems revolving around the equation: $\Delta E = q + w$. The trick with any problem is to be able to translate from one kind of language (words) into another kind of language (equations). If you understand the equation above, then you should have no trouble. There are three variables in this equation: ΔE, q, and w. The equation tells you the relationship between the three terms. So, if you have any two of the three terms, you should be able to determine the third term. It's just simple addition and subtraction. Let's start by calculating ΔE when we are given q and w.

EXERCISE 5.10. A system receives 490 J of heat and delivers 490 J of work to its surroundings. What is the change in the internal energy of the system?

Answer: Recall that $\Delta E = q + w$. Heat and work are the only two ways to transfer energy with the surroundings. The system is gaining 490 J in the form of heat, so q is +490 J. The system is losing 490 J in the form of work, so w = −490 J (don't forget that negative sign).

So $\Delta E = (490\ J) + (−490\ J) = 0\ J$. Therefore, the internal energy does not change. The system has gained exactly as much in the form of heat as it has lost in the form of work

These problems might seem trivial, but it is important to take a few minutes and *do them*. It is like the difference between reading about words in a foreign language, and practicing to speak them. The only way to ever learn a language is to practice speaking it. You can't learn a language by reading a book. Similarly, if you want to become fluent in the language of chemistry, you MUST get practice manipulating equations. Don't be lazy. Do the problems even if they seem simple at first.

PROBLEM 5.11. A system gives 150 calories of heat to the surroundings while delivering 140 calories of work to the surroundings. What is the change in internal energy of the system? Express your answer in kJ. (1 calorie = 4.184 J).

Answer: _____

PROBLEM 5.12. A system receives 0.75 kJ of heat and delivers 0.47 kcal of work. What is the change in internal energy of the system? Express your answer in kJ. [Hint: The units of work and heat are not the same in this problem (kJ vs. kcal), so you must first convert kcal into kJ.]

Answer: _____

In the problems above, we focused on solving for ΔE when we were given q and w. Now, let's just manipulate the equation a little bit:

$$\text{If } \Delta E = q + w$$
$$\text{Then, } q = \Delta E - w$$
$$\text{And } w = \Delta E - q$$

These are just simple manipulations of the equation. So, you should be able to calculate q if you are given ΔE and w. Similarly, you should be able to calculate w if you are given ΔE and q. Let's do some examples:

PROBLEM 5.13. If a system receives 0.54 kJ of heat, how much work must the system do in order for the change in internal energy to be +0.27 kJ?

Answer: _____

PROBLEM 5.14. If a system delivers 0.54 kJ of work to the surroundings, how much energy must be transferred in the form of heat in order for the change in internal energy to be +0.27 kJ?

Answer: _____

Now, let's just make sure you understood the terms *endothermic* and *exothermic*. Recall that exothermic means that q was negative (the system is losing energy in the form of heat to the surroundings), and endothermic means that q is positive (the system is gaining energy in the form of heat from the surroundings). So, if you can figure out whether q is positive or negative, then you will know whether the process is exothermic or endothermic. Let's try a problem:

EXERCISE 5.15. You are performing a chemical reaction. The internal energy of your system does not change at all, but the reaction forms gases and the volume increases. Is the reaction endothermic or exothermic?

Answer: Remember that equations are the language of chemistry. Whenever you have a word problem, it is helpful to rewrite the information in the form of equations. Let's look at the problem with this in mind.

If gas is formed, then the system is doing work on the surroundings. Therefore, the system is losing energy, so w has a negative value. Therefore, the problem can be rephrased like this:

$$\Delta E = 0. \text{ w is a negative number. What is the sign of q?}$$

That's the whole problem. Now that we have rephrased it, the problem becomes simpler. In order to determine if the reaction is endothermic or exothermic, all you need to determine is the sign of q. Is it a positive number or a negative number?

Since $\Delta E = 0$, you get: $0 = q + w$.

Or, $q = -w$.

Therefore, $q = -$(a negative number), which means q must be a positive number. If q is positive, then the system is gaining energy from the surroundings, which means that this is an endothermic process.

PROBLEM 5.16.   You are performing a chemical reaction in which the reactants are gases and the products are solids. Therefore, the volume of the system has decreased. The $P\Delta V$ work is $+0.64$ kJ. The internal energy of your system increases by 0.89 kJ. What is the value of q? Is the reaction endothermic or exothermic?

Answer: _____

PROBLEM 5.17.   You are performing a chemical reaction in which the reactants are gases and the products are solids. Therefore, the volume of the system has decreased. The $P\Delta V$ work is $+1.07$ kJ. The internal energy of your system increases by 0.89 kJ. What is the value of q? Is the reaction endothermic or exothermic?

Answer: _____

5.3 STATE FUNCTIONS AND PATHWAYS

In order to be fluent in the language of thermodynamics, you must understand the difference between state functions and pathways. These terms are crucial and are used all of the time, even when not explicitly stated.

Before translating these terms, let's consider an analogy that will serve as a basis for the definitions. In the previous section, we saw the analogy of the rich uncle who sends you money every month via wire transfers. At any time, you can ask how much money is in your account, because your account balance is *a property of your account*. But you cannot ask how many wire transfers are *in* your account. That makes no sense. A wire transfer is **not** *a property of your account*. Rather, it is one method by which your account balance can interact with a different account somewhere else. If he wanted, your rich uncle could give you the cash instead, and you could deposit the money into the account yourself. That would be a different way of getting to the same point (more money in your account).

Now we can see the difference between a state function and a pathway. A state function is *a property of the system*, like how much money is in your account. It doesn't matter how the money got there. This quantity tells you something about its current state. Examples of state functions are temperature, volume, pressure, and energy. All of these properties are state functions because they each describe a property of the system. By contrast, heat and work are *not* properties of a system. Rather, they are two different pathways that a system can take when going from one state to another.

Although pathways are the way to get from one state to another, the final state of a system does not tell you which pathway it took to get there. Consider the following analogy: Imagine that we have two cups of water. We ask one person to take one cup, walk up to the 100th floor of a tall building, and then to come back down to the 60th floor. We ask a second person to take the second cup and just walk straight up to the 60th floor. At the end of the day, both cups end up next to each other, on the 60th floor. You can do all the measurements in the world, but you will not be able to tell which cup traveled the longer distance. If they both end up on the 60th floor, then they are both the same distance from the ground (probably about 600 feet above ground level). In other words, the height of the cups is not dependent at all on where they *were*—but only on where they *are* right now. That is because the position of an object is a state function. It is a property that is dependent on the current state, not on how it got there. The choice of path only makes a difference on the surroundings (the person who took the cup up 100 floors is probably a bit more tired than the person who took the more direct route).

When you have a system, it is impossible to tell what pathway was taken to get the system to its current state. The only way to tell would be to measure the effects on the surroundings, which is generally too hard to do. So, we must treat state functions as being independent of the pathways they took to get there.

Before we can move on, we must say one more thing about state functions. State functions can either be *extensive* or *intensive*. If you want to be able to correctly solve problems on your exam, you must understand the difference between these two terms. Some properties are not dependent on how much stuff you have in your system. For example, temperature. If you take two cups of water, each at 25° C, and you mix them together in a bowl, experience tells us that the temperature does not rise to 50° C. We know that the temperature will stay at 25° C. Temperature is a property that does not depend on the amount of stuff you have. But what about volume? If the water in the first cup occupied a volume 100 mL, and the water in the second cup also occupied a volume of 100 mL, then mixing them together gives an amount of water that would occupy 200 mL.

So, there are two kinds of properties: those that don't depend on the amount of stuff and those that do. The first category is called *intensive* properties. Examples include temperature and density (does NOT depend on amount of stuff). The second category is called *extensive* properties. Examples include volume and mass (DOES depend on amount of stuff).

The truth is that many students often forget which term is which (intensive and extensive). You can devise your own way to remember which is which. The most important part is that you understand for each state function whether it depends on the amount of stuff or not. When you truly understand a state function, you will be able to clearly say whether it is dependent on the amount of stuff or not, just like we did with temperature and volume.

5.4   WHAT IS ENTHALPY?

Earlier in this chapter, we saw the two ways that energy can be transferred: work and heat. And we expressed it like this:

$$\Delta E = q + w$$

So, if we have some process or reaction, and we want to actually calculate ΔE for the system, then we will need to measure both q and w. It turns out that q is fairly straightforward to measure. To avoid the need to measure w, chemists have defined a new state function, called *enthalpy*. To truly understand why we are introducing this new term, you will need to understand the profound relationship between enthalpy and entropy, and you will probably not learn about that until the second semester of your chemistry course. So, in the meantime, it will be difficult to justify why we need to create this new term called enthalpy. But at the very least, we can try to understand what it is, and that is the goal of this section. To better understand what enthalpy is, let's consider an analogy.

Imagine that you are finished with school, and you get a nice job making $100,000 per year. If I were to ask you how much money you are making, you would have to say, "It depends: before taxes or after taxes?" Your employer might be paying you $100,000 per year, but some of that money is going to the government. It is a lot easier to say how much money you make *before* taxes. That is a simple number, and you agreed to accept your position based on that number. But it is much more difficult to say how much money you are making *after* taxes. That requires a lot of calculations at the end of the year.

In the analogy we just gave, there are two sources of money exchange: (1) you receive money from your employer, and (2) you pay taxes to the government. So, money comes in, and then some money goes out. Now, we can extend this analogy to a system undergoing change. We just need to replace money with energy. Imagine a process where energy is coming in (in the form of heat), and energy is going out (in the form of work). As an example, let's go back to the case of heating a gas in a canister with a moving piston:

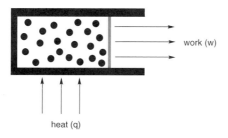

You are giving the gas energy (q) and the gas expands to push on the environment (w). When we compare this to our analogy, the heat coming into the system is like your salary (how much money you make *before* taxes). Then, some energy goes out of the system in the form of work, and that is like the taxes you need to pay. ΔE (which takes into account the energy that comes in as q and the energy that goes out as w) is like the amount of money you make *after* taxes. But if instead of measuring q *and w*, if we just focus on q, then that is like focusing on the amount of money you make *before* taxes. And that is a lot easier to measure. Using our analogy, we will soon see that enthalpy is like the money you make *before* taxes. To understand why, let's take a close look at the way chemists define enthalpy.

It would be convenient to use the letter E for enthalpy, but unfortunately, that letter is already taken (for energy). So we need a new symbol, and we use the letter H. We have defined enthalpy like this:

$$H = E + PV$$

and

$$\Delta H = \Delta E + P\Delta V$$

By defining a new state function that is ΔE plus this extra term ($P\Delta V$), then we have:

$$\begin{aligned} \Delta H &= & \Delta E & + P\Delta V \\ &= & (q + w) & + P\Delta V \\ &= & (q - P\Delta V) & + P\Delta V \\ &= & q \end{aligned}$$

Notice that, in the equations above, P is a fixed value (we say P instead of ΔP, because pressure is not changing). So, the equations tell us that $\mathbf{\Delta H = q}$ under conditions of constant pressure. This fits with the analogy we just gave: the change in enthalpy is like your salary *before* taxes. Your salary just measures how much your employer pays you, and does not take into account how much you have to pay the government. In the same way, ΔH is equal to the energy that came into the system (q), and we are not focusing on the amount of energy that left the system (w). So, you can think of it like this: ΔH is like your salary *before* taxes, and ΔE is like your salary *after* taxes.

But that leaves us with some obvious questions. If ΔH is just q, then why do we need to define a new term called H? Why can't we just use q? What was wrong with q?

In the previous section, we learned about the difference between state functions and pathways. Now we can apply that understanding to help us understand why we need to create a new term called H. We have seen that q represents a *pathway*. q is the ***transfer*** of energy from one place to another (remember our wire-transfer analogy). But H is a *state function*. To see why, look at how we defined it: $H = E + PV$. All of these terms (E, P, and V) are state functions, so H must also be a state function.

Now we can understand why we need this new term. A state function is *not dependent on the path* that was taken to get to the final state. If you have a reaction, it doesn't matter whether the reaction takes place via one step or via twelve steps. All that matters is the initial state and the final state. If we focus on a state function instead of a pathway, we can now talk about the ΔH of a reaction, *regardless of the path from starting materials to products*. Later in this chapter, we will see some important problem-solving techniques that rely heavily on this concept.

Now let's make sure we understand the *units* of enthalpy. We just said that, under conditions of constant pressure, $\Delta H = q$. We already know that q is measured in units of energy (because q is just the transfer of energy). So, the units of ΔH will also be units of energy. ΔH will therefore be measured in Joules or kiloJoules.

While we are talking about the units of enthalpy, there is another important point we should mention. In the previous section, we saw the difference between extensive properties and intensive properties. **Enthalpy is an extensive property**. That means that it is dependent on the amount of stuff that is present. Let's say you look up in a chart that the ΔH for a particular reaction is 50 kJ/mol. If you are running the reaction on 2 moles, then the ΔH will be 100 kJ. If you use the value 50 kJ in your calculations, then you will get the wrong answer. Notice that the units are kJ**/mol.** The **/mol** is an indication that this property is extensive.

Now that we have seen what enthalpy is and how it is measured, let's take a look at some common terminology that chemists use when referring to enthalpy. We have already argued that, under conditions of constant pressure, $\Delta H = q$. Remember that q refers to the transfer of energy in the form of heat. Therefore, we often refer to ΔH as the "heat of reaction". That is a common term used for ΔH, so you should get used to it now.

In general, the ΔH for any reaction is called the heat of reaction, but there are a few specific categories of reactions that get their own special name for ΔH. Here are the two common names that you will see in this course:

1. When one mole of a substance reacts with O_2, the process is a called a *combustion reaction*. The change in enthalpy associated with a combustion reaction is called the **heat of combustion**, and it is shown like this: ΔH_{comb}.

2. When one mole of a substance is formed from its elements, the change in enthalpy associated with the reaction is called the **heat of formation**, and it is shown like this: ΔH_f.

The two examples above are special types of *reactions*. But there are also a couple of special *processes* (not reactions) that also get special names. Here are the two common names that you will see in this course:

1. When one mole of a substance turns from a solid into a liquid (so the solid is melting), there is a change in enthalpy associated with the melting process. This ΔH is called the **heat of fusion**, and it is shown like this: ΔH_{fus}.

2. When one mole of a substance turns from a liquid into a gas (so the liquid is said to vaporize), there is a change in enthalpy associated with the vaporiza-

tion process. This ΔH is called the ***heat of vaporization***, and it is shown like this: ΔH_{vap}.

These two processes both represent phase changes. These are the only two values of ΔH that you will need to describe any phase change. This is shown in the following diagram:

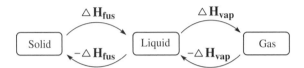

We have now seen several situations that get their own special terms: ΔH_{fus}, ΔH_{vap}, ΔH_{comb}, and ΔH_f. But all of these terms represent the same concept: enthalpy.

There is one more thing we should mention about the way we report enthalpy. The value of ΔH depends on the temperature and pressure at which the process takes place. So, if I say that the enthalpy change for a process is 34 Joules per mole, you should say, "At what temperature and pressure did you do the process?" If you did the process at a different temperature or pressure, the value of ΔH would be different. In order to take this into account, we have created a term called the ***Standard State Enthalpy*** ($\Delta H°$). This means that we are specifically talking about 25° C and 1 atmosphere of pressure. To denote the standard state enthalpy change, we use the degree symbol: $\Delta H°$. You can think of it as reminding you what temperature (and pressure) we are talking about.

5.5   HESS'S LAW

When it comes to calculating enthalpies, there are two common types of problems that you need to know how to do. The skills involved in doing these problems are really part of the basic language of chemistry, and therefore, we will devote the rest of this chapter to these skills. Both types of problems are based on Hess's Law, so we need to start by taking a close look at Hess's Law. The key to understanding Hess's Law is to remember that enthalpy is a state function.

Recall the difference between state functions and pathways. A state function is a property of the system, and is NOT dependent on the path that was taken to get to that state. To see how this can be useful in calculating enthalpies, let's go back to an analogy we saw earlier in this chapter.

We used the analogy of two cups of water being carried to the 60th floor of a tall building via two different routes. We said that the height of the cup of water (how high it was above street level) was dependent only on where it was (it doesn't matter how it got there). So, if one cup goes up to the 100th floor and then back down to the 60th floor, it will be at the same height as another cup that started on the ground floor and just went straight up to the 60th floor. The path taken to get to the final state is irrelevant, because position is a state function.

Now let's extend that analogy a bit to help us understand Hess's Law. Let's say you are given the following information: (1) you have to go up 990 feet to get from the 1st floor to the 100th floor, and (2) you have to go down 400 feet to get from the 100th floor to the 60th floor. Let's express this information in a diagram:

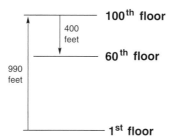

With this information, you should be able to calculate the change in height if you went *directly* from the 1st floor to the 60th floor:

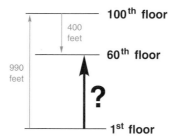

Although this is a different path (a more direct path) to get to the 60th floor, it is still the same final position. And it is clear to see how we would calculate the difference in height between the 1st floor and the 60th floor. It would just be (990 ft − 400 ft). So the 60th floor is 590 feet above the 1st floor. These calculations only work because position is a state function. Your position is not dependent on the path you took to get there. If you can calculate the change in height on one path (going up to the 100th floor and then back down), then you will automatically know the change in height along any other path that gets you to the same position.

This kind of calculation will also work with any other state function. So, instead of talking about position (or height), we will use the same trick for enthalpy, which is also a state function. If we want to calculate the change in enthalpy to go from one state to another state, then we can choose any path that takes us from the initial state to the final state. Consider the diagram below:

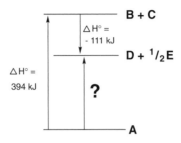

There is no difference between this diagram and the one shown earlier when we were talking about height. It is exactly the same concept. Only this time, we are talking about the reaction of A being converted into D and E. And instead of measuring a change in height, we are measuring a change in enthalpy. Since enthalpy is a state function, we can choose a more roundabout path to get to the final state. We can first turn A into B and C, and then turn B and C into D and E. As long as we get there in the end, it doesn't matter if we take this longer path or the more direct path. Whether we go up and then back down, or whether we go directly from the initial to the final state, the change in enthalpy should be the same along either path. We can calculate that $\Delta H°$ for the long path is 394 kJ − 111 kJ = 283 kJ, so the short path must also have $\Delta H° = 283$ kJ.

This trick is called Hess's Law, and it can be used to calculate enthalpy changes for chemical reactions. An example that can be found in many textbooks is the formation of CO_2 from carbon monoxide and oxygen. This is what the diagram for that looks like:

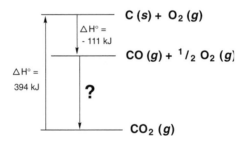

Notice that this diagram is almost exactly the same as the previous diagram, but now we want to form CO_2, rather than break it apart into carbon and oxygen. In our analogy, we would say that we want to go from the 60th floor back down to the 1st floor, rather than going from the 1st floor up to the 60th floor. So, in this case, $\Delta H° = -283$ kJ (notice the minus sign, as compared to the last example).

It makes sense, but students often get confused, because many problems involve reactions that must first be manipulated before you can just add up the enthalpy changes. So, you need to learn how to manipulate reactions for problems like this one:

You are given the value of $\Delta H°$ for each of the following two reactions:

$$N_2O_5 \text{ (s)} \longrightarrow 2NO \text{ (g)} + \tfrac{3}{2}O_2 \text{ (g)} \qquad \Delta H° = 224 \text{ kJ}$$
$$NO \text{ (g)} + \tfrac{1}{2}O_2 \text{ (g)} \longrightarrow NO_2 \text{ (g)} \qquad \Delta H° = -57 \text{ kJ}$$

Using this information, calculate the value of $\Delta H°$ for the following reaction:

$$2NO_2 \text{ (g)} + \tfrac{1}{2}O_2 \text{ (g)} \longrightarrow N_2O_5 \text{ (s)} \qquad \Delta H° = ?$$

Now we're not talking about simple diagrams anymore. Perhaps what really confuses students is all of the numbers. When you see something like $2NO_2$ (g), it is hard to know what to focus on. There is a coefficient, a compound with two different elements, different subscripts, and the phase is shown as well. This can be overwhelming. But you can ignore the phase, and you can treat the entire compound as a single unit. So, in the beginning we will not use specific compounds (like N_2O_5 or CO_2 or SO_3, etc.). Rather, we will use A, B, C, etc. to refer to compounds so that we can focus on building skills and not getting lost in the chemical formulas. Once we have developed the essential skills, then we will bring the formulas back in, and you will see that the method for solving the problem does not change at all. You just have to know what to look for when solving these problems.

Let's start by learning how to add two reactions together to get a third reaction.

EXERCISE 5.19. You are given the value of $\Delta H°$ for each of the following two reactions:

$$A + B \longrightarrow C \qquad \Delta H° = 104 \text{ kJ}$$
$$A + C \longrightarrow D \qquad \Delta H° = -34 \text{ kJ}$$

Using this information, calculate the value of $\Delta H°$ for the following reaction:

$$2A + B \longrightarrow D \qquad \Delta H° = ?$$

Answer: This problem is asking you to add the first two reactions above to get the third reaction. The way to do this is to rewrite the first two reactions. Then, underneath them, add everything on the left sides of the two reactions, and then do the same for the right side, like this:

$$A + B \longrightarrow C \qquad \Delta H° = 104 \text{ kJ}$$
$$\underline{A + C \longrightarrow D \qquad \Delta H° = -34 \text{ kJ}}$$
$$2A + B + C \longrightarrow C + D \quad \Delta H° = (104 \text{ kJ}) + (-34 \text{ kJ})$$

Now, you will notice that there is a C on each side of your sum total reaction, so you can cross off the C on either side. When you do that, you get:

$$2A + B \longrightarrow D$$

and that is the reaction for which you are trying to calculate the value of $\Delta H°$. So, that means that if you do the first two reactions, one after another, then you will get to the same exact place as if you had just done the one reaction you care about. Remember what you learned before about Hess's Law: If the first two reactions (added together) get you to the same place in the end as the third reaction does, then $\Delta H°$ of the third reaction must be equal to the sum of the $\Delta H°$ values from the first two reactions. When you add those values together, you get: $(104 \text{ kJ}) + (-34 \text{ kJ}) = 70$ kJ. And that's your answer.

Notice that you did not need to manipulate the reactions at all. You just added them together and got your third reaction. Let's practice a few more problems like this, and then we will work on problems where we will need to do some manipulation.

PROBLEM 5.20.   You are given the value of $\Delta H°$ for each of the following two reactions:

$$2A + B \longrightarrow 2C \qquad \Delta H° = -57 \text{ kJ}$$

$$2C + A \longrightarrow D \qquad \Delta H° = 78 \text{ kJ}$$

Using this information, calculate the value of $\Delta H°$ for the following reaction:

$$3A + B \longrightarrow D \qquad \Delta H° = ?$$

Let's do this problem step-by-step. Start by rewriting the first two reactions, and adding them together here:

1.

2. ————————————————————

   sum total:

Notice that you have 2C on each side of the sum total of the first two reactions, so cross them off, and you should see that this is the reaction for which you are trying to calculate $\Delta H°$. When you add the $\Delta H°$ values from the first two reactions, you get ——————.

PROBLEM 5.21.   You are given the value of $\Delta H°$ for each of the following two reactions:

$$A \longrightarrow B + C \qquad \Delta H° = -92 \text{ kJ}$$

$$C \longrightarrow B + D \qquad \Delta H° = -64 \text{ kJ}$$

Using this information, calculate the value of $\Delta H°$ for the following reaction:

$$A \longrightarrow 2B + D \qquad \Delta H° = ?$$

Try to do this one on your own. Use the space below.

Now that you have seen the basic idea, you need to learn how to manipulate reactions. The truth is that there are only two basic ways that reactions can be manipulated. You can *reverse* a reaction, or you can *multiply* the whole reaction by a number. Let's start with the first type of manipulation: reversing a reaction.

Here is the rule: when you reverse a reaction, you must change the sign of $\Delta H°$. For example, if

$$A \longrightarrow B + C \qquad \Delta H° = -92 \text{ kJ}$$

then

$$B + C \longrightarrow A \qquad \Delta H° = +92 \text{ kJ}$$

Here we reversed the reaction, so we reversed the sign of $\Delta H°$ from -92 kJ to $+92$ kJ. Let's see an example of where this is used:

EXERCISE 5.22. You are given the value of $\Delta H°$ for each of the following two reactions:

$$C + D \longrightarrow A \qquad \Delta H° = -188 \text{ kJ}$$
$$\frac{1}{2}C + D \longrightarrow B \qquad \Delta H° = -286 \text{ kJ}$$

Using this information, calculate the value of $\Delta H°$ for the following reaction:

$$A \longrightarrow B + \frac{1}{2}C \qquad \Delta H° = \text{?}$$

Answer: You can clearly see that adding the first two reactions is not going to get you the third reaction, because the first reaction has A on the right side, and the third reaction has A on the left side. So, before you add the first two reactions together, you must reverse the first reaction. DON'T FORGET to change the sign of $\Delta H°$. Reverse the first reaction and then rewrite the reactions:

$$A \longrightarrow C + D \qquad \Delta H° = +188 \text{ kJ}$$
$$\frac{1}{2}C + D \longrightarrow B \qquad \Delta H° = -286 \text{ kJ}$$

Notice that you changed the sign of $\Delta H°$ in the first reaction to $+188$ kJ. Now you are ready to add the two reactions together. When you do, you get:

$$A + \frac{1}{2}C + D \longrightarrow C + D + B$$

Now, you must cross off anything that appears on both sides of this sum total reaction. There is a D on both sides, so if you cross them off, you get:

$$A + \tfrac{1}{2}C \longrightarrow C + B$$

Notice that you also have a C on both sides, but you can't just cross them both off, because the left side has only half of a mole of C and the right side has a full mole of C. So, cross off half of a mole of C from each side. That gets rid of any traces of C on the left side, and it leaves half a mole of C on the right side:

$$A \longrightarrow B + \tfrac{1}{2}C$$

And that is the reaction for which you need to calculate $\Delta H°$. So, you have: $\Delta H° = +188$ kJ $+ (-286$ kJ$) = -98$kJ. Remember that you had to use $+188$ kJ instead of -188 kJ because you reversed the first reaction. One of the most common mistakes in these types of problems is to forget to reverse the sign of $\Delta H°$ when you reverse a reaction.

Now let's get some practice:

PROBLEM 5.23.   You are given the value of $\Delta H°$ for each of the following two reactions:

$$A \longrightarrow B \qquad \Delta H° = 152 \text{ kJ}$$
$$\tfrac{1}{2}B \longrightarrow C \qquad \Delta H° = -58 \text{ kJ}$$

Using this information, calculate the value of $\Delta H°$ for the following reaction:

$$A + C \longrightarrow \tfrac{3}{2}B \qquad \Delta H° = ?$$

(Hint: Notice that the first reaction has A on the left side, which is perfect; but the second reaction has C on the right side, which is not good.)

Let's do one more practice problem where you have to reverse one of the reactions:

PROBLEM 5.24.   You are given the value of $\Delta H°$ for each of the following two reactions:

$$2A + B \longrightarrow 2C \qquad \Delta H° = -310 \text{ kJ}$$
$$2A + \tfrac{1}{2}B \longrightarrow D \qquad \Delta H° = -169 \text{ kJ}$$

Using this information, calculate the value of $\Delta H°$ for the following reaction:

$$D + \tfrac{1}{2}B \longrightarrow 2C \qquad \Delta H° = ?$$

Now, we are ready to see the only other manipulation that you need to know how to do. If you **multiply** the entire reaction by some coefficient, then you must multiply the value of $\Delta H°$ by the same coefficient (remember that enthalpy is an extensive property). For example, if

$$A \longrightarrow B + C \qquad \Delta H° = 100 \text{ kJ}$$

then

$$2A \longrightarrow 2B + 2C \qquad \Delta H° = 200 \text{ kJ}$$

Let's see a problem that uses this manipulation:

EXERCISE 5.25. You are given the value of $\Delta H°$ for each of the following two reactions:

$$A + \tfrac{1}{2}E \longrightarrow C \qquad \Delta H° = -283 \text{ kJ}$$

$$D + E \longrightarrow 2B \qquad \Delta H° = -180 \text{ kJ}$$

Using this information, calculate the value of $\Delta H°$ for the following reaction:

$$A + B \longrightarrow C + \tfrac{1}{2}D \qquad \Delta H° = ?$$

Answer: Start by looking to see if you need to do the first kind of manipulation (reversing the reaction). You can see that the first reaction has A on the left side, and that is where it needs to be in the end, so the first reaction does not need to be reversed. The second reaction has B on the right side, but it needs to be on the left side in the end, so you must reverse the second reaction. By doing that, you get:

$$A + \tfrac{1}{2}E \longrightarrow C \qquad \Delta H° = -283 \text{ kJ}$$

$$2B \longrightarrow D + E \qquad \Delta H° = +180 \text{ kJ}$$

Notice that you changed the sign of $\Delta H°$ for the second reaction. Now every compound is on the correct side of each reaction, but these two reactions still do not add up to give the correct reaction. The reaction you want has B, not 2B. Whenever you see that the coefficients don't add up properly to get the desired reaction, then that should be a signal to you that you need to multiply one of the reactions by some number.

In this case, you have 2B, but the coefficient should be 1 instead of 2. So, you need to multiply your second equation by the number $\tfrac{1}{2}$ (in other words, you must divide your second reaction by two). When you do that, you need to multiply every coefficient in the reaction by $\tfrac{1}{2}$, and that gives you:

$$A + \tfrac{1}{2}E \longrightarrow C \qquad \Delta H° = -283 \text{ kJ}$$

$$B \longrightarrow \tfrac{1}{2}D + \tfrac{1}{2}E \qquad \Delta H° = +90 \text{ kJ}$$

Notice that the value of ΔH° for the second reaction was multiplied by ½ as well, to give 90 kJ. Whenever you multiply a reaction by a number, you must multiply ΔH by that number as well.

Now when you add the reactions (and cross off the ½E on either side), the sum total is exactly the reaction you are looking for. So ΔH° = (−283 kJ) + (90 kJ) = −193 kJ.

Now, try to get some practice on the next two problems. Remember that the order goes like this: first look at each reaction to decide if it needs to be reversed (and if so, don't forget to reverse the sign of ΔH°), and then, look at each reaction to see if it needs to be multiplied by a coefficient (and if so, don't forget to multiply ΔH° by the same coefficient). That's it. Just two manipulations that you need to know how to do.

Once you get used to doing these manipulations with A, B, C, etc., it is actually the exact same thing with real compounds. Let's finish up practicing to make sure you know how to do it, and then we can do some problems with real compounds.

PROBLEM 5.26. You are given the value of ΔH° for each of the following two reactions:

$$A + B \longrightarrow C \qquad \Delta H° = -95 \text{ kJ}$$
$$D \longrightarrow \tfrac{1}{2}A + \tfrac{1}{2}C \qquad \Delta H° = -90 \text{ kJ}$$

Using this information, calculate the value of ΔH° for the following reaction:

$$2A + B \longrightarrow 2D \qquad \Delta H° = ?$$

PROBLEM 5.27. You are given the value of ΔH° for each of the following two reactions:

$$A \longrightarrow \tfrac{1}{2}B + C \qquad \Delta H° = -75 \text{ kJ}$$
$$\tfrac{1}{2}B \longrightarrow \tfrac{1}{2}C + D \qquad \Delta H° = +105 \text{ kJ}$$

Using this information, calculate the value of ΔH° for the following reaction:

$$A + 2D \longrightarrow \tfrac{3}{2}B \qquad \Delta H° = ?$$

Now we are ready to do some problems with real compounds, instead of A, B, and C. The trick is to treat all compounds like a single entity. If you see $2N_2O_5$, then treat it in your mind as just 2A, where A = N_2O_5. Don't get bogged down by all of the subscripts in each compound. You need to learn to ignore them and instead focus on the information necessary to solve the problem. In these types of problems, you should focus on two things: the coefficients and the number of different chemical entities present. Let's see how to do this in the following example:

EXERCISE 5.28. You are given the value of $\Delta H°$ for each of the following two reactions:

$$H_2 \ (g) + O_2 \ (g) \longrightarrow H_2O_2 \ (l) \qquad \Delta H° = -188 \ kJ$$

$$H_2 \ (g) + \tfrac{1}{2}O_2 \ (g) \longrightarrow H_2O \ (l) \qquad \Delta H° = -286 \ kJ$$

Using this information, calculate the value of $\Delta H°$ for the following reaction:

$$H_2O_2 \ (l) \longrightarrow H_2O \ (l) + \tfrac{1}{2}O_2 \ (g) \qquad \Delta H° = \ ?$$

Answer: For now, don't get confused about what phase each reagent is in (gas, liquid, or solid). We will deal with this in a little bit. Instead, focus on the fact that this problem is not different in any way from what you have done before.

In the final reaction (for which you need to calculate ΔH), you see that H_2O_2 is on the left side of the reaction. Now look for this compound in the reactions you are given, and you will find it in the first reaction. But is it on the right side? No— so you must reverse the first reaction. Notice that you are doing this problem just like you have done all of the other problems. The only difference is that you are dealing with a real compound (H_2O_2) instead of the letter A. But the method for solving the problem is identical.

Now, try to go through the rest of this problem yourself. If you are having trouble dealing with real compounds instead of A, B, and C, then you can always rewrite the problem in the form that makes it more comfortable for you. To do that, you would just go through each reaction and look at how many different compounds are present in all of the reactions. In the two reactions given above, there are a total of four different compounds. You assign a letter to each one (H_2 = A, O_2 = B, H_2O_2 = C, and H_2O = D), and then you can rewrite the equations using A, B, C, and D instead of the real compounds. If that makes you more comfortable, then you can do that. But you must eventually get to the point where you can solve a problem without the need to rewrite the equations. There is nothing inherently wrong with rewriting the equations, but it just wastes precious time on an exam, so you want to get to a point where you can treat each compound as a single entity, without assigning a letter to it.

You've actually done this exact problem already. Give it a try, and when you think you have the answer, go back and take a look at Exercise 5.22 to check your answer.

Now let's try one more problem that you have also already done before. Once again, try to do this problem *without* rewriting the equations using A, B, C, etc. If you find that you are getting stuck, then you can resort to rewriting the equations.

PROBLEM 5.29.   You are given the value of $\Delta H°$ for each of the following two reactions:

$$2Cu\ (s) + O_2\ (g) \longrightarrow 2CuO\ (s) \qquad \Delta H° = -310 \text{ kJ}$$
$$2Cu\ (s) + \tfrac{1}{2}O_2\ (g) \longrightarrow Cu_2O\ (s) \qquad \Delta H° = -169 \text{ kJ}$$

Using this information, calculate the value of $\Delta H°$ for the following reaction:

$$Cu_2O\ (s) + \tfrac{1}{2}O_2\ (g) \longrightarrow 2CuO\ (s) \qquad \Delta H° = ?$$

If you get stuck, go back and look at Problem 5.24.

Let's review real quickly what we have seen. When calculating $\Delta H°$ for a reaction, based on known $\Delta H°$ values for two other reactions, you have to do a few things:

1. Decide if any of the reactions need to be reversed (if so, reverse the sign of $\Delta H°$).

2. Decide if any of the reactions need to be multiplied by a coefficient (if so, multiply $\Delta H°$ by the same coefficient).

3. Add the reactions together to give a sum total reaction, and cross off anything present on both sides.

4. Make sure this reaction is the one you are looking for. If so, just add the values of $\Delta H°$ to get your answer.

Now that you know the procedure, you should realize that you can do the exact same type of problem when you have three reactions to add instead of just two reactions. Everything is the same. For each reaction, you have to decide if it needs to be reversed or multiplied by a coefficient. But when you have three reactions, it can be a bit trickier. In some situations, you will find that it is difficult to decide whether or not to reverse one of the reactions. Let's do an example, and we will see where it gets tricky:

EXERCISE 5.30.   You are given the value of $\Delta H°$ for each of the following three reactions:

$$H_2 (g) + \tfrac{1}{2}O_2 (g) \longrightarrow H_2O (l) \qquad \Delta H° = -289 \text{ kJ}$$
$$N_2O_5 (g) + H_2O (l) \longrightarrow 2HNO_3 (l) \qquad \Delta H° = -77 \text{ kJ}$$
$$\tfrac{1}{2}N_2 (g) + \tfrac{3}{2}O_2 (g) + \tfrac{1}{2}H_2 (g) \longrightarrow HNO_3 (l) \qquad \Delta H° = -174 \text{ kJ}$$

Using this information, calculate the value of $\Delta H°$ for the following reaction:

$$N_2 (g) + \tfrac{5}{2} O_2 (g) \longrightarrow N_2O_5 (g) \qquad \Delta H° = ?$$

Answer: Begin by reversing all of the reactions that need to be reversed, and then move on to any multiplying that you need to do. In order to determine which reactions need to be reversed, you should notice that the desired reaction (for which you need to calculate $\Delta H°$) has three compounds: N_2, O_2, and N_2O_5. Each one of these three compounds is present in one or more of the three reactions above. So, you just need to look for these compounds in the three reactions, and you need to determine if they are on the correct side of each reaction.

Let's start with the product (N_2O_5). It only appears in the second reaction. And it appears on the wrong side (on the left side instead of the right side). So, you know that you will need to reverse the second reaction.

Now let's look for the other two compounds (N_2 and O_2). N_2 can only be found in the third reaction, and it is on the correct side. So you don't need to reverse the third reaction.

Now comes the tricky part: O_2. This compound is in two reactions. It is in the first reaction and the third reaction. You've already decided that you are not going to reverse the third reaction, but what about the first reaction? It would seem like you should reverse it to get the O_2 on the left side where it needs to be. BUT, maybe if you leave it alone, it will get crossed off by the O_2 in the third reaction. So, do you reverse the reaction or not? The best thing to do at this point is to rewrite the reactions (reversing whatever you decided needs to be reversed), and then look for some clues. In this example, you've decided to reverse the second reaction and not the third, so you get:

$$H_2 \, (g) + \tfrac{1}{2}O_2 \, (g) \longrightarrow H_2O \, (l) \qquad \Delta H° = -289 \text{ kJ}$$

$$2HNO_3 \, (l) \longrightarrow N_2O_5 \, (g) + H_2O \, (l) \qquad \Delta H° = +77 \text{ kJ}$$

$$\tfrac{1}{2}N_2 \, (g) + \tfrac{3}{2}O_2 \, (g) + \tfrac{1}{2}H_2 \, (g) \longrightarrow HNO_3 \, (l) \qquad \Delta H° = -174 \text{ kJ}$$

Now you need to look for clues to see whether or not to reverse the first reaction. Always begin looking at the compounds that must cross off completely because they are not present at all in the desired reaction. You'll notice that H_2O is present in the first and second reactions, but it is not present in the desired reaction. That tells you that H_2O must cross off completely when you add the reactions. The way it is right now, they won't cross each other off because H_2O is on the right side of both reactions. Therefore, one of the reactions must be reversed. You've already decided that the second reaction must be the way it is so that N_2O_5 will be on the correct side. So, you can conclude that the first reaction must be reversed:

$$H_2O \, (l) \longrightarrow H_2 \, (g) + \tfrac{1}{2}O_2 \, (g) \qquad \Delta H° = +289 \text{ kJ}$$

$$2HNO_3 \, (l) \longrightarrow N_2O_5 \, (g) + H_2O \, (l) \qquad \Delta H° = +77 \text{ kJ}$$

$$\tfrac{1}{2}N_2 \, (g) + \tfrac{3}{2}O_2 \, (g) + \tfrac{1}{2}H_2 \, (g) \longrightarrow HNO_3 \, (l) \qquad \Delta H° = -174 \text{ kJ}$$

So, in the end, you reversed two reactions. Now you are ready to multiply. In order to get these reactions to add up properly, you need to multiply the third reaction by the number 2 (which means that you will also need to multiply $\Delta H°$ of that reaction by 2):

$$H_2O \, (l) \longrightarrow H_2 \, (g) + \tfrac{1}{2}O_2 \, (g) \qquad \Delta H° = +289 \text{ kJ}$$

$$2HNO_3 \, (l) \longrightarrow N_2O_5 \, (g) + H_2O \, (l) \qquad \Delta H° = +77 \text{ kJ}$$

$$N_2 \, (g) + 3 \, O_2 \, (g) + H_2 \, (g) \longrightarrow 2 \, HNO_3 \, (l) \qquad \Delta H° = -348 \text{ kJ}$$

These three reactions will now add up to give your desired reaction. So, $\Delta H°$ of the desired reaction will be:

$$\Delta H° = (+289 \text{ kJ}) + (+77 \text{ kJ}) + 2(-174 \text{ kJ}) = \mathbf{18 \text{ kJ}}$$

Now, let's try to get some practice on a couple of problems where you are given three reactions. Remember that you should first look for the compounds that are present in the desired reaction. Find those compounds in the three reactions, and determine whether they are on the correct side. One of the reactions might be tricky. If that happens, look for the compounds that are not present in the desired reaction, and keep in mind that those compounds must cross off in the end. That will probably help. Then multiply by any of the reactions necessary to have all three reactions add up to the desired the reaction.

PROBLEM 5.31. You are given the value of $\Delta H°$ for each of the following three reactions:

$$H_2S\ (g) \longrightarrow H_2\ (g) + S\ (s) \qquad \Delta H° = 20.36\ kJ$$
$$S\ (s) + O_2\ (g) \longrightarrow SO_2\ (g) \qquad \Delta H° = -296.83\ kJ$$
$$H_2O\ (l) \longrightarrow H_2\ (g) + \tfrac{1}{2}O_2\ (g) \qquad \Delta H° = 285.83\ kJ$$

Using this information, calculate the value of $\Delta H°$ for the following reaction:

$$2H_2S\ (g) + SO_2\ (g) \longrightarrow 3\ S\ (s) + 2\ H_2O\ (l) \qquad \Delta H° = ?$$

PROBLEM 5.32. You are given the value of $\Delta H°$ for each of the following three reactions:

$$2Cu\ (s) + S\ (s) \longrightarrow Cu_2S\ (s) \qquad \Delta H° = -79.5\ kJ$$
$$S\ (s) + O_2\ (g) \longrightarrow SO_2\ (g) \qquad \Delta H° = -296.8\ kJ$$
$$Cu_2S\ (s) + 2O_2\ (g) \longrightarrow 2CuO\ (s) + SO_2\ (g) \qquad \Delta H° = -527.5\ kJ$$

Using this information, calculate the value of $\Delta H°$ for the following reaction:

$$Cu\ (s) + \tfrac{1}{2}O_2\ (g) \longrightarrow CuO\ (s) \qquad \Delta H° = ?$$

There is one thing that should be kept in mind that hasn't been mentioned yet, because we were dealing with A, B, and C instead of real compounds for most of the skill-building exercises. When you are working on a problem and you have gotten to the point where you have added the reactions to get a sum total reaction, you can only cross off something that appears on both sides *in the same physical state* (solid, liquid, or gas). If you have A (g) on one side and A (l) on the other side, you CANNOT cross them off. They are not the same. There is actually a $\Delta H°$ involved for the transformation from one to the other. You will rarely encounter this, but you should keep it in mind. In a case where you do encounter this, the transformation of A (g) to A (l) would be considered as a separate reaction that would need to be added to give the sum total reaction.

Now, you should be ready to practice more problems in your textbook. Good luck. . . .

5.6 ENTHALPY OF FORMATION

In the previous section, you saw situations where you were given the $\Delta H°$ values for two or more reactions, and you had to manipulate those reactions to find $\Delta H°$ for another reaction. But there is another common way to calculate $\Delta H°$ for a reaction. To see how it works, let's revisit the Lego analogy we saw earlier in this book. Let's say that you have a Lego structure that is already built, and you want to take it apart and build a different structure using the exact same materials:

Notice that the final structure has all of the same pieces as the initial structure. In other words, this is a balanced reaction.

Now we will combine this Lego analogy with Hess's Law to understand the second important way to calculate enthalpy changes for reactions. Recall that Hess's Law tells us that we can take any path to get to the products, and the total enthalpy change should be the same as if we took any other path. So, let's say we want to calculate the enthalpy change for a reaction, for example:

$$H_2SO_4 \longrightarrow SO_3 + H_2O \qquad \Delta H° = \,?$$

If we know $\Delta H°$ for breaking apart H_2SO_4 into its elements (hydrogen, sulfur, and oxygen), and if we know $\Delta H°$ for recombining those elements to form SO_3 and H_2O, then we have all the information we need in order to calculate $\Delta H°$ for the reaction above. Once again, we are using Hess's Law, which tells us that *any path* that gets us from reactants to products will have the same $\Delta H°$ as any other path that gets us to the end result. The information that we have been given can be shown in the following diagram:

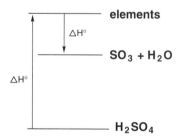

Based on this diagram, we can see how we would calculate $\Delta H°$ for our reaction. It doesn't matter that we will not *actually* break apart H_2SO_4 into its elements. But as long as we can calculate $\Delta H°$ for this *hypothetical* pathway, then we will know $\Delta H°$ for our reaction.

Before we see how to do this, we should mention an important (and often confusing) point. We are not talking about breaking apart the compound into its individual atoms. We are talking about breaking up the compound into its elements *in their standard states*. A standard state refers to the form in which the element is found in nature at 25° C and 1 atm pressure. For example, oxygen exists as a diatomic molecule (O_2) in its standard state. Under these conditions, oxygen does not exist as the atom O, but rather as O_2. So, if you have a compound with three oxygen atoms, then breaking up the compound into its elements will not produce 3 O, but rather $\frac{3}{2} O_2$ (which is essentially still three oxygen atoms, just expressed in terms of O_2 rather than O). In the Lego analogy given earlier, we would now express it like this:

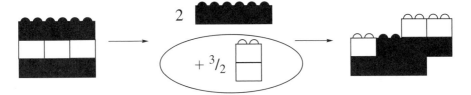

So the idea goes like this: you break apart the reactants into their elements in their standard states, and then you recombine them to form the products. Here's how you do it. You can take any compound and ask what the $\Delta H°$ would be if you formed that compound from its simple elements in their standard states. For example, you can form one mole of liquid water (H_2O) from H_2 and $\frac{1}{2}O_2$, and there is an enthalpy change associated with this *formation* of water from its simple elements in their standard states. Therefore, we call it the *standard enthalpy of formation*, and we represent it with a subscript "f" like this: $\Delta H°_f$. There are actually long lists of $\Delta H°_f$ values for many common compounds. If you know the value of $\Delta H°_f$ for each compound in a balanced reaction, then you have all the information you need in order to calculate $\Delta H°$ for that reaction.

As an example, let's consider the reaction we discussed earlier:

$$H_2SO_4 \longrightarrow SO_3 + H_2O \qquad \Delta H° = ?$$

If we know the $\Delta H°_f$ values for all of the compounds above (H_2SO_4, SO_3, and H_2O), then we can calculate $\Delta H°$ of the reaction above. Let's take a close look at how we do this, and we'll start by looking at the products. The first product we are forming is SO_3, and in order to do this, we will need one sulfur atom and three oxygen atoms (but remember that the standard state of oxygen is O_2), so we can express it like this:

$$H_2SO_4 \longrightarrow \boxed{\begin{array}{c} SO_3 \\ \uparrow {\scriptstyle \Delta H°_f (SO_3)} \\ S + \frac{3}{2}O_2 \end{array}} + H_2O$$

The ΔH associated with the formation of SO_3 from its elements in their standard states is called $\Delta H°_f(SO_3)$. Now, let's consider our second product, H_2O. In order to form H_2O, you will need two hydrogen atoms (H_2) and one oxygen atom ($\frac{1}{2}O_2$). And there is a ΔH associated with the formation of water from its elements in their standard states. It is called $\Delta H°_f(H_2O)$:

Now, we can focus on the reactants. In this case, there is only one: H_2SO_4. Let's imagine *breaking apart* this compound into its elements:

Notice that the elements we get when we break up H_2SO_4 are exactly the same elements that we need *to form our two products.* That should make sense, because we started with a balanced reaction. So the question now is: what is the ΔH associated with *breaking apart* H_2SO_4 into its elements? Recall from the previous section that anytime you reverse a reaction, you must change the sign of ΔH. And that is all we need to do here. If the ΔH associated with *forming* H_2SO_4 is called $\Delta H°_f$ (H_2SO_4), then the ΔH associated with *breaking apart* H_2SO_4 must be $-\Delta H°_f$ (H_2SO_4). Notice the negative sign. To break apart H_2SO_4 is just the opposite of forming H_2SO_4:

$$H_2 + S + 2O_2 \longrightarrow H_2SO_4 \qquad \Delta H°_f\ (H_2SO_4)$$
$$H_2SO_4 \longrightarrow H_2 + S + 2O_2 \qquad -\Delta H°_f\ (H_2SO_4)$$

So, we can now place this value into our diagram:

And, now we have built up a diagram that shows us an alternate path to get to our products. We just use $-\Delta H°_f (H_2SO_4)$ to break up our reactant into its elements, and then we use $\Delta H°_f (SO_3)$ and $\Delta H°_f (H_2O)$ to reassemble those same elements into our products. So, the $\Delta H°$ associated with our reaction will be the sum of all of these individual $\Delta H°_f$ values:

$$\Delta H° = \Delta H°_f (SO_3) + \Delta H°_f (H_2O) - \Delta H°_f (H_2SO_4)$$

Recall that in the last section, we saw *two* rules for manipulating reactions: (1) reversing reactions and (2) multiplying coefficients. Well, in our discussion in this section, we have just taken into account the reversing of reactions (the reactants have a negative sign in front of their $\Delta H°_f$ values). But what about taking coefficients into account?

This becomes important when there are coefficients in your reaction. For example:

$$2A + 3B \longrightarrow 3C \qquad \Delta H° = ?$$

In this example, we can't just use the formula $\Delta H°_f (C) - \Delta H°_f (A) - \Delta H°_f (B)$. We are actually forming more than 1 mole of C. Since we are forming 3 moles of C, we need to use the value $\underline{3\Delta H°_f} (C)$. Similarly, we need to use $-2\Delta H°_f (A)$, because we are breaking apart two moles of A. And we need to use $-3\Delta H°_f (B)$, because we are breaking apart 3 moles of B.

So, we see that $\Delta H°$ for any reaction is just the $\Delta H°_f$ values of the products minus the $\Delta H°_f$ values of the reactants, taking coefficients into account. We can express this generically as shown here:

For the reaction: aA + bB $\longrightarrow$ cC + dD

a, b, c and *d* are the coefficients that balance the reaction; A, B, C, and D are the chemical entities; and we find that:

$$\Delta H° = c\Delta H°_f (C) + d\Delta H°_f (D) - a\Delta H°_f (A) - b\Delta H°_f (B)$$

Let's get some practice with this concept:

EXERCISE 5.33. For the following reaction, write a formula expressing $\Delta H°$ of the reaction in terms of $\Delta H°_f$ of each of the compounds:

$$2A + \tfrac{3}{2}B \longrightarrow \tfrac{3}{2}C + 2D$$

Answer: We said that $\Delta H°$ for any reaction is just the $\Delta H°_f$ values of the products minus the $\Delta H°_f$ values of the reactants. So, remember to put a negative sign in front of the $\Delta H°_f$ of both A and B. Also, don't forget to take the coefficients into account. So, you get:

$$\Delta H° = \tfrac{3}{2}\Delta H°_f (C) + 2\Delta H°_f (D) - 2\Delta H°_f (A) - \tfrac{3}{2}\Delta H°_f (B)$$

Now let's do some problems:

PROBLEM 5.34. For the following reaction, write a formula expressing $\Delta H°$ of the reaction in terms of $\Delta H°_f$ of each of the compounds:

$$A + \tfrac{1}{2}B \longrightarrow \tfrac{1}{2}C$$

Answer: $\Delta H° = $ _____

PROBLEM 5.35. For the following reaction, write a formula expressing $\Delta H°$ of the reaction in terms of $\Delta H°_f$ of each of the compounds:

$$3A + B \longrightarrow 2C + \tfrac{1}{2}D$$

Answer: $\Delta H° = $ _____

There is just one more subtle point to address before we can do some real examples. When we are calculating $\Delta H°$ for a reaction, and the reaction involves a reactant or product that is an element in its standard state (like O_2), then the $\Delta H°_f$ for that element will just be zero. So, when we are constructing our formula to calculate $\Delta H°$ for the overall reaction, we can just ignore the $\Delta H°_f$ of the element. To help understand why, consider the following question: what is the $\Delta H°_f$ of O_2 (g)? It is the $\Delta H°$ associated with making O_2 (g) from O_2 (g). And obviously, that must be zero. When we calculate $\Delta H°$ for a reaction, like in the examples above, we are *adding* $\Delta H°_f$ of each *product,* and we are *subtracting* $\Delta H°_f$ of each *reactant.* Therefore, any $\Delta H°_f$ of zero can be ignored.

Let's see an example of this:

EXERCISE 5.36. Consider the following reaction:

$$2NO\ (g) + O_2\ (g) \longrightarrow N_2O_4\ (g)$$

Calculate $\Delta H°$ based on the following data:

$$\Delta H°_f \text{ of NO} = +90.4 \text{ kJ/mol}$$
$$\Delta H°_f \text{ of } N_2O_4 = +9.16 \text{ kJ/mol}$$

Answer: We said that $\Delta H°$ for any reaction is just the $\Delta H°_f$ values of the products minus the $\Delta H°_f$ values of the reactants. Don't forget to take the coefficients into account. So, you get:

$$\Delta H° = (1 \text{ mol}) \times \Delta H°_f\ [N_2O_4\ (g)] - (2 \text{ mol}) \times \Delta H°_f\ [NO\ (g)]$$
$$- (1 \text{ mol}) \times \Delta H°_f\ [O_2\ (g)]$$
$$= (1 \text{ mol}) \times (9.16 \text{ kJ/mol}) - (2 \text{ mol}) \times (90.4 \text{ kJ/mol})$$
$$- (1 \text{ mol}) \times (\textit{0 kJ/mol})$$
$$= -171.6 \text{ kJ}$$

Notice that $\Delta H°_f$ of O_2 is just zero. You can see this clearly if you draw a diagram similar to the one drawn for the previous example:

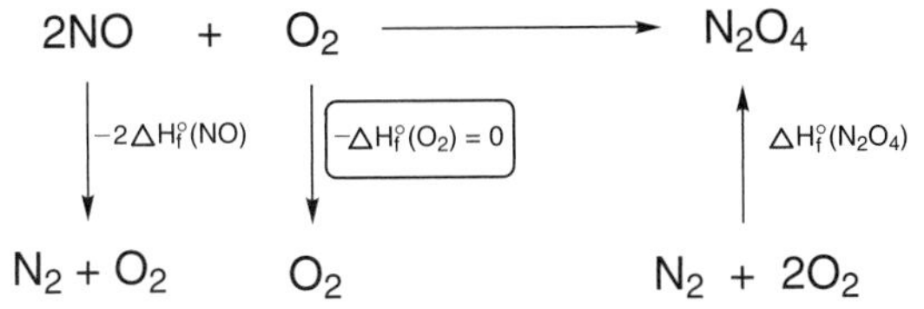

$$2NO \quad + \quad O_2 \xrightarrow{\hspace{3cm}} N_2O_4$$

$\Big\downarrow -2\Delta H_f^\circ(NO) \qquad \boxed{\Big\downarrow -\Delta H_f^\circ(O_2) = 0} \qquad \Big\uparrow \Delta H_f^\circ(N_2O_4)$

$$N_2 + O_2 \qquad\quad O_2 \qquad\qquad\qquad N_2 + 2O_2$$

So, you could have just ignored this, and it would not have affected your answer.

PROBLEM 5.37. Consider the following reaction:

$$2\ NH_3\ (g) + 3\ N_2O\ (g) \xrightarrow{\hspace{2cm}} 4\ N_2\ (g) + 3\ H_2O\ (l)$$

Calculate $\Delta H°$ based on the following data:

$$\Delta H°_f \text{ of } NH_3\ (g) = -46.0 \text{ kJ/mol}$$
$$\Delta H°_f \text{ of } N_2O\ (g) = +81.5 \text{ kJ/mol}$$
$$\Delta H°_f \text{ of } H_2O\ (l) = -285.9 \text{ kJ/mol}$$

PROBLEM 5.38. Consider the following reaction:

$$C_2H_4\ (g) + O_3\ (g) \xrightarrow{\hspace{2cm}} CH_3CHO\ (g) + O_2\ (g)$$

Calculate $\Delta H°$ based on the following data:

$$\Delta H°_f \text{ of } C_2H_4\ (g) = +51.9 \text{ kJ/mol}$$
$$\Delta H°_f \text{ of } O_3\ (g) = +143 \text{ kJ/mol}$$
$$\Delta H°_f \text{ of } CH_3CHO\ (g) = -167 \text{ kJ/mol}$$

PROBLEM 5.39. Consider the following reaction:

$$2 SO_2 (g) + O_2 (g) \longrightarrow 2 SO_3 (g)$$

Calculate $\Delta H°$ based on the following data:

$$\Delta H°_f \text{ of } SO_2 (g) = -297 \text{ kJ/mol}$$
$$\Delta H°_f \text{ of } SO_3 (g) = -396 \text{ kJ/mol}$$

PROBLEM 5.40. Consider the following reaction:

$$3 NO_2 (g) + H_2O (l) \longrightarrow 2 HNO_3 (l) + NO (g)$$

Calculate $\Delta H°$ based on the following data:

$$\Delta H°_f \text{ of } NO_2 (g) = 34 \text{ kJ/mol}$$
$$\Delta H°_f \text{ of } H_2O (l) = -285.9 \text{ kJ/mol}$$
$$\Delta H°_f \text{ of } HNO_3 (l) = -174.1 \text{ kJ/mol}$$
$$\Delta H°_f \text{ of } NO (g) = +90.4 \text{ kJ/mol}$$

CHAPTER **6**

ORBITALS, BONDS, AND COUNTING ELECTRONS

When electrons overlap with each other, they can form bonds. So if we want to understand what a bond is, we will first need to understand a few things about electrons. The first half of this chapter will be mostly theory and analogies, while the second half of the chapter will focus on the calculations associated with counting electrons.

6.1 ATOMIC ORBITALS

There are two models that can be used to describe what an electron is. The first model treats an electron as a small particle circling around the nucleus of an atom (in the same way that planets orbit our sun). In fact, you probably learned this model at some point in time. The second model treats an electron as a wave that follows a path around the nucleus. The problem is that neither of these models does a perfect job in explaining *all* of the properties of electrons. Some properties can best be explained by treating an electron as a particle. Other properties can best be explained by treating an electron as a wave. So, what is an electron? A particle or wave?

Actually, it behaves like both. That doesn't mean that it changes its mind from one second to the next. It doesn't flip back and forth every other second. It is what it is, and unfortunately, we don't have one model that will explain everything. The way that we deal with this is to meld the two models together into one model, and we do this mathematically. A wave function is a mathematical equation that describes electrons as having wave-like properties *and* particle-like properties simultaneously. But does that really help us form an image in our minds of what electrons are? Not really. What does a particle-wave look like?

The mathematicians and physicists tell us that we should not try to picture what an electron looks like in our minds. Any attempt to do this will end up limiting our understanding of what it really is. In order to truly understand what an electron is, we must give up on trying to imagine what it looks like, and instead, we must focus on the elegant mathematics that describe it. Now, that might work well for a physicist or a mathematician, but it doesn't work for me. And I am guessing it doesn't work for you either. We naturally want to come up with a picture of it, so that we can wrap our hands around it. So, let's go against the advice of the physicists and the mathematicians, and try to paint a picture that will help us visualize what an electron is.

Keep in mind that any analogy will be imperfect, because we simply don't have any macroscopic objects that behave like electrons. But let's use the follow-

169

ing analogy, with the appreciation that it will not be a perfect analogy. Let's treat electrons as if they are *clouds*; just like clouds in the sky. A cloud can be thicker in the center, and thinner at the edges. In fact, the edges of a cloud are not so well defined. It is hard to say where the cloud exactly ends. It just tapers off gradually.

Now imagine that there is a cloud in the sky that is shaped like a balloon (and it is three-dimensional). Once again, the edges are somewhat fuzzy, because it is a cloud. The cloud is very thick in the center, and it gets thinner as you move away from the center. Clouds can also come in different shapes and sizes. We can imagine a cloud shaped like a basketball, and many other shapes as well. That is the analogy that we will use to think about electrons.

Before we continue, we should point out some of the key limitations of this analogy, and that will actually help us to better understand electrons.

- Real clouds can come in any shape or size. Electron clouds only come in a small number of shapes and sizes. We call these clouds *orbitals*, which is a pretty bad name because it reminds us of the incorrect model where electrons are flying around the nucleus like planets around a sun (if it were up to me, I would call them clouds instead of orbitals).

- A real cloud is made of many individual water molecules. But an electron cloud is not made of billions of particles. You have to think of it as *one entity*. The electron cloud can be thicker in some places and thinner in other places, but it is *not* made up of many small little particles.

- Just like real clouds have fuzzy edges, so do electron clouds. A real cloud tapers off in thickness, but at some point, you can clearly see that the cloud has ended. In contrast, electron clouds *never* end. When we talk about orbitals, we are considering the region of space that contains 95% of the electron cloud. As you move farther and farther away, the cloud gets thinner and thinner, but it never stops. In fact, believe it or not, if you want to consider the region of space that contains 100% of any electron cloud, you will need to consider the entire universe.

We said before that orbitals don't come in any shape or size. All atoms have very defined orbitals that can be filled by electrons. Each orbital can hold a maximum of two electrons. There are s orbitals, p orbitals, d orbitals, and f orbitals. All of these orbitals are called **atomic orbitals,** because they are regions of space *within one atom*. In your textbook, you will have pictures of these orbitals, and you will learn about which orbitals are in each energy level of an atom. This is pretty straightforward stuff. The part that can be confusing is how these atomic orbitals are used to form bonds. And that is what we will focus on now.

6.2 MOLECULAR ORBITALS

Bonds are formed when an atomic orbital of one atom overlaps with an atomic orbital of another atom. To help us understand how atomic orbitals can overlap, let's revisit our cloud analogy. Imagine that it is a clear, sunny day, and you look up to see two clouds in the sky. The clouds are approaching each other, and at some point, they begin to overlap with each other and occupy some of the same space. This picture can help us understand what it means for two orbitals to overlap, but unfortunately, that's where the analogy breaks down. When two clouds partially overlap, you can still see the regions that are not overlapping. However, when two atomic orbitals overlap, their nature changes so drastically that they don't look like they did before they overlapped.

To understand what happens to the orbitals when they overlap, we will use a new analogy. Imagine that there are two people living in two neighboring houses; one person in each house. One day, they decide that it would be much better to live together in a mansion than to live separately in two modest-sized homes. So, they perform some magic, and a tornado comes. The tornado pulls apart both of their homes, and all of the raw materials (bricks, wood, glass, etc.) go swirling up in the air. Then the tornado spits back all of the raw material onto the ground, to form two new structures. One is a mansion, and the other is a small out-house:

The materials that make up the two new structures are all exactly the same; they are just distributed differently than before. If you look at the new scenery, you would see that there are still two houses. But now, the two houses are very different from the original houses: one is a mansion and the other is an out-house. These new houses do not look anything like the original two houses. And here comes the important part: the two original houses *do not exist anymore*. The tornado destroyed them and turned them into two new houses.

Now that we have new houses, both people can live comfortably in the mansion, while the out-house remains vacant. It would be very hard to convince one of the two people to go live in the out-house, as that would be very uncomfortable. Now let's try to capture all of this information in a diagram that will show that the two original houses were destroyed to form two new houses, AND our diagram will also show the relative amount of discomfort associated with living in each type of house:

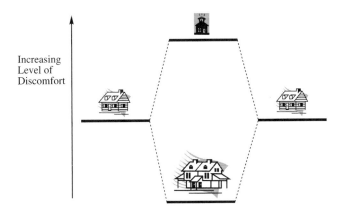

Increasing
Level of
Discomfort

This diagram shows us the two original houses (that don't exist anymore) and the two new houses, so that we can compare the level of discomfort associated with living in each type of house. We can see that the most uncomfortable house to live in is the out-house. The most comfortable house is the mansion. And the two original houses used to be mediocre. Now we can understand why the two people would want to do this whole magic show in the first place. It made their living accommodations more comfortable.

Electrons behave in a similar way. Before a bond is formed between two atoms, you have an electron in each atom "living" alone. Each electron is in an atomic orbital by itself (an orbital can be thought of as an electron's house, except that orbitals don't have solid walls the way houses do). When the two atomic orbitals overlap with each other, they *cease to exist*, and two new orbitals appear. One of these new orbitals is like a mansion (which is very comfortable, or very *low in energy*), and both electrons go into that new low-energy orbital. The high-energy orbital (the out-house) is left vacant:

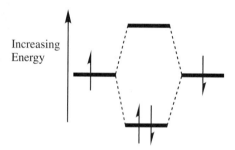

In this diagram, we see the original two orbitals, each containing one electron to begin with. Then, these orbitals combine to form two *new* orbitals, and both electrons go into the lower energy orbital. And that is the definition of a bond.

These two new orbitals can no longer be called *atomic* orbitals, because neither of the new orbitals is centered on just one atom. Rather, these new orbitals are centered *in between* two atoms (and in some cases, the orbitals can be spread out over an even larger region of a molecule). So, we call them *molecular orbitals*. In our analogy, atomic orbitals are the two original houses, and molecular orbitals are the mansion and the out-house. The mansion is called the *bonding molecular orbital*, because the *bond* is formed when the two electrons lower their energy by going into this molecular orbital. The out-house is called the *anti-bonding molecular orbital*, because the two electrons would be higher in energy (less stable) if they went into this molecular orbital.

In our analogy, it might be possible to pay one of the people enough money to go live in the out-house for one night. Similarly, it is possible to give one of the electrons enough energy to "jump" from the bonding molecular orbital into the anti-bonding molecular orbital (for a brief moment). There is an entire field of science devoted to how, when, and why that happens. It is called spectroscopy, which is discussed in more detail in an organic chemistry course.

For now, it is important to understand the energy diagram just shown. You will see many more of these diagrams in this course, and you will need to know how to read them. That is why we spent so much time on the analogy.

Now that we understand what a bond is, let's turn our attention to important types of bonds.

6.3 ELECTRONEGATIVITY, INDUCTION, AND POLARITY

In the previous section, we saw how two atoms can share electrons to form a bond. But we cannot assume that the two atoms are sharing the electrons equally. To see what we mean by this, we will need to revisit our cloud analogy.

We said in the beginning of this chapter that we should not think of electrons as *particles* that are in orbit of some nucleus. Also, we should not think of them as waves surrounding a nucleus. Rather, they are some combination of these two models. The best way to visualize that is to think of an electron as being a ***cloud of electron density***. These clouds have shapes (which we call orbitals). They can be thick in some places, and they can be thin in other places. So, any atom can have a ***partial*** negative charge ($\delta-$) or a ***partial*** positive charge ($\delta+$). There is one very important way that these charges can arise, and it is called induction. Induction will be the basis for calculating oxidation states later in this chapter, so it is critical that we understand it. To explain induction, we really need to start with *electronegativity*.

The term *electronegativity* is used to measure the relative tendency, or "desire", of an atom to attract electron density toward itself. You probably remember this term from your high school chemistry class. When two atoms form a bond between them, each of the atoms will pull on the electron density that is being shared between them. If the atoms have different electronegativity values, then each atom pulls on the electron density with a different strength. The atom that is more electronegative will pull harder, and this will cause $\delta-$ and $\delta+$ to form. This is called induction, and it is generally shown with a special type of arrow:

Since fluorine is more electronegative than carbon, fluorine will pull harder on the electron density. The arrow shows which way the electron density is being pulled, and the tail of the arrow almost looks like a positive charge to show you where the $\delta+$ is.

For any bond, you should be able to draw the arrow in the correct direction if you know the electronegativity values for both atoms. To determine which atom is more electronegative, you do ***not*** need to memorize values. You just need to have access to a periodic table, and you need to know the following trends for electronegativity:

Increasing electronegativity

H																	
Li	Be												B	C	N	O	F
Na	Mg												Al	Si	P	S	Cl
K	Ca	Sc	Ti	V	Cr	Mn	Fe	Co	Ni	Cu	Zn	Ga	Ge	As	Se	Br
Rb	Sr	Y	Zr	Nb	Mo	Tc	Ru	Rh	Pd	Ag	Cd	In	Sn	Sb	Te	I
Cs	Ba	La	Hf	Ta	W	Re	Os	Ir	Pt	Au	Hg	Tl	Pb	Bi	Po	At

Increasing electronegativity

The arrows above indicate increasing electronegativity. So, fluorine is the most electronegative atom. Notice that we have not included the noble gases (the last column of the periodic table). These elements typically do not form bonds with other atoms because they already have all the electrons they need, so we will not discuss their electronegativity.

Hydrogen is one important exception to the trends shown in the chart above. The electronegativity of H falls in between B and C. That is the only one that you need to remember, because it will not be obvious when you look at a periodic table.

With the trends above, it is easy to compare two atoms in the same *row* of the periodic table. For example, let's compare C and N. You don't actually need to know the exact electronegativity values. You just look at the periodic table, and you can see that nitrogen is farther to the right than carbon, so nitrogen is more electronegative than carbon. Therefore, if you have a bond between carbon and nitrogen, the inductive effect looks like this:

Similarly, it is also easy to compare two atoms in the same *column* of the periodic table. For example, if we compare F and Cl, we find that F is higher up on the periodic table, so a bond between F and Cl will have the following inductive effect:

$$\overleftarrow{\underset{F-Cl}{\qquad}}$$

It is also common to compare two atoms that are not in the same row or the same column. For example, Mg and Br are not in the same column or the same row. But Mg is all the way on the left side of the periodic table, and Br is all the way on the right side. So, Br will definitely pull harder.

$$\overrightarrow{\underset{Mg-Br}{\qquad}}$$

There are a few situations where it is hard to say which element is more electronegative. For example, compare O and Cl. Look at the periodic table, and find both of them. If you use the trends we discussed, you will see that you cannot tell which element is more electronegative. In these situations, you would need to actually look up the electronegativity values in your textbook in order to determine the direction of the inductive effect. These situations, luckily, are pretty rare. So, we can rely on the trends for most situations. (It might be helpful to know that oxygen is the second most electronegative atom, after fluorine).

EXERCISE 6.1. Identify the direction of the inductive effect in a C-O bond.

Answer: Look at the periodic table. You'll see that C and O are in the same row of the periodic table. O is farther to the right on the periodic table, so O is more electronegative than C. Therefore, the inductive effect goes like this:

$$\overset{\xrightarrow{\hspace{1.5cm}}}{C-O}$$

For each of the following bonds, identify the direction of the inductive effect using an arrow:

6.2. Na—O   **6.3.** B—C   **6.4.** S—O

6.5. N—O   **6.6.** B—I   **6.7.** Mg—O

6.8. Si—C   **6.9.** O—H   **6.10.** Cl—F

We will soon see that the ability to determine the direction of the inductive effect is critical for calculating oxidation states. You will find that a familiarity with the periodic table will allow you to calculate oxidation states more quickly. The truth is that you will rarely see many of the elements on the periodic table, so you don't need to be familiar with the entire periodic table. Rather, you should focus on the elements that are commonly found in problems throughout your textbook:

H																	
Li												B	C	N	O	F
Na	Mg											Al	Si	P	S	Cl
K	Ca	Sc	Ti		Cr	Mn	Fe	Co	Ni	Cu	Zn					Br
								Pd	Ag		Sn				I
								Pt	Au	Hg	Pb				

Spend a few moments reviewing the positions of the elements shown above. These are the elements that you will see again and again.

6.4 COVALENT AND IONIC BONDS (H-BONDS)

Now that we have seen how to determine the direction of the inductive effect for any bond, we can turn our attention to the magnitude of the inductive effect. Consider the following two examples:

$$H_3C\text{—}OH \qquad Li\text{—}OH$$

In each case above, the oxygen is the more electronegative element. But, the oxygen is not pulling on the electron density equally in these two cases. To see why, think of the inductive effect as a "tug-of-war". When C and O share a bond, there is a tug-of-war between them, and O is pulling stronger than C:

When Li and O share a bond, there is also a tug-of-war, but there is no competition. Li is on the left side of the periodic table, and O is on the right side. The O is so much more electronegative than Li, and O is pulling so much harder than Li. So, we no longer think of these atoms as *sharing* the electrons. Rather, we treat it as if the O has stolen the electrons completely:

This kind of bond is called an ionic bond (because the electrons are not really *shared* anymore, and the force of the bond stems from the force of attraction between the two ions). The opposite extreme is when electrons are being shared perfectly, as in the following example:

$$H_3C\text{—}CH_3$$

Since C and C have the same electronegativity, the electrons in between the two carbon atoms are being equally shared. This type of bond is called a covalent bond.

But what about the case we saw before (C—O)? The difference in electronegativity between C and O is not great enough for this bond to be ionic. But, it is also not purely covalent, because the C and O do not have the same electronegativity. So, we call this bond a polar covalent bond.

Now we have two extremes (covalent bonds and ionic bonds), and we have the kind of bond that is somewhere in between these two extremes (polar covalent). The truth is that most bonds in molecules will fall into the last category (polar covalent). But you should know how to recognize the different kinds of bonds when you see them.

To determine whether a bond is covalent, polar covalent, or ionic, you just need to look at the periodic table and see how far apart the elements are. If they are on opposite sides of the periodic table (like Li-O), then you can be sure the bond is ionic. If they are fairly close to each other (like C-O), then you can be sure the bond is polar covalent. But what about cases where you are not sure if the elements are far enough apart for the bond to be ionic? Is there some dividing line that can help us make a decision? For example, it would be helpful to have a rule like this: if the difference in electronegativity is greater than X, then we call it an ionic bond.

Some textbooks will give this kind of rule of thumb. But there does not seem to be good agreement about where to put this "dividing line". Some textbooks might say that a bond is ionic if the difference in electronegativity is **greater than 2**. Other textbooks place the "dividing line" **at 1.7**. Most textbooks do not give a dividing line at all. So, we will focus on bonds where you can quickly determine whether a bond is ionic or polar covalent by looking at the proximity of the elements on the periodic table. Let's see an example:

EXERCISE 6.11. Determine whether the following bond is covalent, polar covalent, or ionic: S—O.

Answer: Look at the periodic table, and locate S and O:

H																
Li	Be											B	C	N	**O**	F
Na	Mg											Al	Si	P	**S**	Cl
K	Ca	Sc	Ti	V	Cr	Mn	Fe	Co	Ni	Cu	Zn	Ga	Ge	As	Se	Br
Rb	Sr	Y	Zr	Nb	Mo	Tc	Ru	Rh	Pd	Ag	Cd	In	Sn	Sb	Te	I
Cs	Ba	La	Hf	Ta	W	Re	Os	Ir	Pt	Au	Hg	Tl	Pb	Bi	Po	At

S and O are very close to each other on the periodic table, so this bond is polar covalent.

There is one exception to this approach. The electronegativity of hydrogen is somewhere in between that of B and C. So when you are applying the method above (looking at the proximity of the elements on the periodic table), you should treat hydrogen *as if* it is located between B and C:

													H					He
Li	Be											B	C	N	O	F	Ne
Na	Mg											Al	Si	P	S	Cl	Ar
K	Ca	Sc	Ti	V	Cr	Mn	Fe	Co	Ni	Cu	Zn	Ga	Ge	As	Se	Br	Kr
Rb	Sr	Y	Zr	Nb	Mo	Tc	Ru	Rh	Pd	Ag	Cd	In	Sn	Sb	Te	I	Xe
Cs	Ba	La	Hf	Ta	W	Re	Os	Ir	Pt	Au	Hg	Tl	Pb	Bi	Po	At	Rn

In fact, the electronegativity of H is so close to the electronegativity of B and C, that we treat B—H and C—H bonds as if they are polar covalent.

For each of the following bonds, determine whether the bond is covalent, polar covalent, or ionic: (Remember that the electronegativity of H falls somewhere in between B and C.)

6.12. Na—O   **6.13.** C—H   **6.14.** O—O

6.15. Li—H   **6.16.** O—F   **6.17.** Mg—O

6.18. Li—F   **6.19.** O—N   **6.20.** B—H

While we are on the topic of types of bonds, we should quickly discuss hydrogen bonds. It is actually a mistake to think of these as bonds. The naming is actually somewhat deceiving. A hydrogen bond is really just an interaction that can take place between two molecules (or between two different regions of one molecule). This interaction is fairly weak when compared to the strength of a bond, and a hydrogen bond is therefore not strong enough to hold the two molecules together permanently.

For example, two water molecules can form a hydrogen bond with each other. This means that when they approach each other in space, there is a force of attraction that pulls them together for a brief instant, but then they continue on their own independent ways. You will learn about forces of attraction in your textbook when you read the chapter covering the different kinds of forces between molecules.

I mention this now because students seem to get confused, with good reason, by the terminology that chemists use to describe this interaction. So the obvious question is: why do chemists call it a hydrogen **bond** if it is not really a bond?

It is true that one hydrogen "bond" is not strong enough to hold two molecules together permanently. But imagine that we have two molecules that can form millions of hydrogen bonds between each other. The collective strength of all of the individual hydrogen bonds can be more than enough to hold the two molecules together. And that is exactly what is happening in DNA. There are two strands coiled around each other in a double helix, and these two strands are held together by millions of hydrogen bonds in between them (forming something that looks like a twisted ladder, where each rung of the ladder is a hydrogen bond). In this double helix, the two strands are held together quite nicely by the collective strength of all of the hydrogen bonds, so we use the term "bonds" to refer to the interactions holding the strands together. This actually explains how the strands can be so easily separated from each other (like a zipper being slowly unzipped).

The bottom line of this discussion is: don't think of a *hydrogen bond* as being a type of bond. It is really just a type of *interaction*, and it belongs in the discussion of forces between molecules (called intermolecular forces).

6.5 FORMAL CHARGES AND OXIDATION STATES

In the previous section, we saw how to predict the presence of partial charges ($\delta-$ and $\delta+$) that are produced by inductive effects. Chemists are actually able to quantify the amount of $\delta-$ or $\delta+$ at any location, but this requires very sophisticated equations and techniques. You will *not* be expected to learn these equations or techniques. But, you should be able to give a *rough estimate* of the electron density at any atom, and there are *two* ways to do this. These two ways are called: *formal charge* and *oxidation state*.

To determine the formal charge or oxidation state of an atom, you must consider *all of the bonds* the atom has. This emphasis on bonds should make sense, because bonds represent the sharing of *electron density*. In the previous section, we saw two extreme situations for bonds: covalent and ionic.

$$H_3C - CH_3 \qquad\qquad H_3C^{\ominus} \quad {}^{\oplus}Li$$
$$\textit{Covalent} \qquad\qquad\qquad \textit{Ionic}$$

We have already argued that most bonds fall somewhere in between these two extremes. But, considering these two extremes can be very useful, because they give us two important ways to determine a rough estimate of the electron density at any atom. We can assume *all bonds* are *covalent* (even though most are not), or we can assume all bonds are *ionic* (even though most are not).

The first approach (assuming all bonds to be *covalent*) is the way we measure the *formal charge* of an atom. The second approach (assuming all bonds to be *ionic*) is the way we measure the *oxidation state* of an atom. Since both approaches are based on extreme assumptions, *neither* approach will provide us with an exact value for the electron density at an atom. We use these approaches because they provide us with two quick ways to get a *rough estimate* of the electron density at any atom. We will soon see that chemists use both of these methods regularly, so we will need to know how to assign a formal charge *and* an oxidation state for any atom in a molecule. To summarize:

1. Assume all bonds are *covalent* → *formal charge.*

2. Assume all bonds are *ionic* → *oxidation state.*

In both of these methods, we count how many electrons the atom is *supposed* to have, and we compare that to the number of electrons the atom *actually* has. The difference between the two methods is how we count the number of electrons an atom actually has. Since both methods start by counting how many electrons an atom is *supposed* to have, let's start there.

To count how many electrons an atom is *supposed* to have, you will need to look at the periodic table. Let's say you are calculating the formal charge or oxidation state of an oxygen atom in some molecule. Either way, your first step is to find O on the periodic table, and ask what column O is in. This will tell you how many valence electrons O is supposed to have (valence electrons are the electrons in the valence shell, or the outermost shell of electrons). Oxygen is in the *sixth* column of the periodic table (column 6A), so it is supposed to have *six* valence electrons.

Once you have determined how many electrons the atom is *supposed* to have, you need to determine how many electrons it *actually* has in the specific molecule you are inspecting. This is where the two methods differ. Let's start with formal charge.

If you are calculating the formal charge, then you imagine that every bond is covalent, even though this assumption is not true. So, we assume that all electrons are being shared equally. As an example, let's consider a C-O bond. This bond is really polar covalent, but we will treat it as if it is covalent. So, we count the electrons as if one is on the carbon atom and the other is on the oxygen atom:

Treat this bond like

C—O $\Longrightarrow$ C• •O

a *covalent bond*

Then, we do this same procedure for all bonds in the molecule. To see how we do this, let's consider the central carbon atom below:

Treat all of these bonds like *covalent bonds*

Now, we count how many electrons are on the carbon atom, and we see that there are four. This is the way we count the electrons when we are determining formal charges.

The last step is to compare the number of electrons C is supposed to have, and the number of electrons C actually has. C is supposed to have four electrons because it is in the fourth column of the periodic table. And in this case, C actually has four electrons. So it has the correct number of electrons. Therefore, there is no formal charge. If C actually had five electrons, then it would have had a formal charge of -1. If we had counted that C only had three electrons, then it would have been missing an electron, and we would have given C a formal charge of $+1$.

In many cases, you will also have to consider lone pairs. A lone pair is when you have two electrons in an atomic orbital, and they are not being used for bonding. For example, consider the structure of ammonia:

Lone Pair

H–N–H
   |
   H

Nitrogen is in the fifth column of the periodic table, so it has five valence electrons. Three of these electrons are used to form bonds, and the remaining two electrons are called a lone pair (we will see much more on this in Chapter 7). When we are counting how many electrons the nitrogen atom actually has, the lone pair counts as two electrons. Let's see an example where lone pairs are involved:

EXERCISE 6.21. Consider the structure of the hydroxide ion, and identify what the formal charge is on the oxygen atom.

$$:\ddot{O}-H$$

Answer: To calculate the formal charge on the oxygen atom, begin by looking at the periodic table to determine how many electrons oxygen is supposed to have. You'll see that oxygen is in the sixth column, so it is *supposed* to have six electrons. Next, count how many electrons it *actually* has in this case, and you do that by treating all bonds as being covalent. In this case, there is only one bond, so you treat it as if the oxygen atom has one electron and the hydrogen atom has one electron:

$$\boxed{:\ddot{O}\cdot}\ \cdot H$$

When you count the number of electrons around the oxygen atom, you find that there are seven (notice that each lone pair counts as two electrons because those electrons are not being shared with any other atom). So, in this case, oxygen has one more electron than it is supposed to have. It should only have six electrons, but it has seven. Since it has one extra electron (an extra negative charge), it will have a formal charge of -1. We show that by placing a minus sign in a circle next to the oxygen atom:

$$\overset{\ominus}{:}\ddot{O}-H$$

For each of the following compounds, identify the formal charge (if any) on the central atom:

6.22.
$$\begin{array}{c} H \\ | \\ H-N-H \\ | \\ H \end{array}$$

6.23. $Cl-Al-Cl$ with Cl below

6.24. $H-\ddot{N}-H$

6.25. $\begin{array}{c} H-\ddot{O}-H \\ | \\ H \end{array}$

6.26.
$$\begin{array}{c} H \\ | \\ H-B-H \\ | \\ H \end{array}$$

6.27. $H-\ddot{S}-H$

Now let's turn our attention to calculating *oxidation states* (which are sometimes referred to as *oxidation numbers*). We said before that the procedure is very similar to the procedure we used in calculating formal charges. The difference is that we treat all bonds as being *ionic* (rather than covalent). Let's take a closer look.

Except for bonds between two atoms of the same element (like C-C), we imagine that every bond is ionic, even though most bonds are not. We assume that one atom has both electrons, and the other atom has none. The atom that is more electronegative gets both electrons. As an example, let's consider a C-O bond again (we looked at this bond earlier as our example when we talked about formal charges). This bond is really polar covalent, but now we will treat it as if it is ionic. Since oxygen is more electronegative than carbon, we count the electrons as if both are on the oxygen atom:

Then, we do this same procedure for all bonds in the molecule. To see how we do this, let's consider the central carbon atom below (this is the same example we used when we saw how to calculate formal charges). We won't draw in charges right now, so that it will be easier to just focus on the electrons and count them all up:

Now, we count how many electrons are on the carbon atom, and we see that there are four. And that is how we count the electrons when we are calculating oxidation states.

The last step is to compare the number of electrons C is supposed to have, and the number of electrons C actually has in this example. C is supposed to have four electrons because it is in the fourth column of the periodic table. And in this case, C actually has four electrons. So it has the correct number of electrons. Therefore, the oxidation state is 0. If C actually had five electrons, then it would have had an oxidation state of -1. If we had counted that C only had three electrons, then it would have been missing an electron, and we would have given C an oxidation state of $+1$.

Once again, don't forget to count lone pairs. They count as two electrons, just as they did when we were calculating formal charges.

EXERCISE 6.28. Determine the oxidation state of the carbon atom in formic acid:

$$\overset{\cdot\cdot\overset{\cdot\cdot}{O}\cdot}{\underset{}{\overset{\|}{H-C-\overset{\cdot\cdot}{\underset{\cdot\cdot}{O}}H}}}$$

Answer: To calculate the oxidation on the carbon atom, begin by looking at the periodic table to determine how many electrons carbon is *supposed* to have. You'll see that carbon is in the fourth column, so it is supposed to have four electrons. Next, count how many electrons it *actually* has in this case, and you do that by treating all bonds as being ionic. Carbon is only slightly more electronegative than hydrogen, but nevertheless, treat it as an ionic bond, and give both electrons to carbon. The remaining bonds are with oxygen atoms, so give all of those electrons to the oxygen atoms. The carbon atom will not get any of those electrons:

$$\overset{\cdot\cdot\overset{\cdot}{O}}{\underset{\cdot\cdot}{}}$$

$$H \quad \boxed{:C} \quad :OH$$

When you count the number of electrons around the carbon atom, you'll find that there are only two electrons. So, in this case, carbon is missing two electrons. It should have four electrons, but it only has two. Since it is missing two negatively charged electrons, it will have an oxidation state of +2.

For each of the following compounds, identify the oxidation state of the central atom:

6.29. $\overset{\cdot\cdot}{\underset{H}{H-N-H}}$ **6.30.** $\underset{Cl}{Cl-Al-Cl}$ **6.31.** $\underset{\cdot\cdot}{H-\overset{\cdot\cdot}{O}-H}$

6.32. $\overset{\cdot O\cdot}{\underset{}{\overset{\|}{H-C-H}}}$ **6.33.** $\underset{H}{\overset{H}{H-C-H}}$ **6.34.** $Cl-Mg-Cl$

For each of the following compounds, identify the oxidation state of each carbon atom:

6.35. $\overset{\cdot\cdot\overset{\cdot}{O}\cdot}{\underset{H}{\overset{\|}{H-C-C-H}}}$ **6.36.** $\overset{\cdot\cdot\overset{\cdot}{O}\cdot}{\overset{\|}{H-C-O-H}}$ **6.37.** $\overset{H}{\underset{H}{\overset{|}{:O:}\atop\overset{|}{H-C-H}}}$

   Although oxidation states and formal charges are useful tools, they are sometimes misleading. In many cases, **neither** the oxidation state **nor** the formal charge is a good way to estimate electron density. As an example, consider the compound we saw in our discussions on formal charge and oxidation states:

$$
\begin{array}{c}
\mathrm{OH} \\
|\\
\mathrm{H_3C-C-H} \\
|\\
\mathrm{CH_3}
\end{array}
$$

We saw that the central carbon atom has an oxidation state of 0. We also saw that it has no formal charge. So, both of these methods imply that the carbon atom is neither $\delta+$ nor $\delta-$. But, the carbon atom actually is fairly $\delta+$, because of the strong inductive effect from the oxygen:

$$
\begin{array}{c}
\uparrow\mathrm{OH} \\
|\\
\mathrm{H_3C} \overset{+}{-} \mathrm{C-H} \\
|\\
\mathrm{CH_3}
\end{array}
$$

The method for assigning formal charges does *not* take this into account. The method for assigning an oxidation state *does* take it into account, *but* it also takes into account that carbon is more electronegative than hydrogen. It treats these two effects as being the same magnitude, so they cancel each other out. But in fact, they are not the same magnitude. This example illustrates that our two methods are really just rough estimates. In some cases, it is more appropriate to talk about formal charges, while in other cases, it may be more appropriate to talk about oxidation states.

In general, oxidation states are more important in a general chemistry course, and formal charges are more important in an organic chemistry course. But that doesn't mean that you should ignore formal charges altogether, because there is an important use for formal charges that we will see in Chapter 7 on drawing Lewis structures.

We will soon see why oxidation states are so important in general chemistry, but first, we need to see how to calculate oxidation states when it's not so easy to just count up the electrons.

6.6 CALCULATING OXIDATION STATES FROM A MOLECULAR FORMULA

In the previous section, we saw how to calculate the oxidation state of any atom in a compound when we are given a drawing that shows us where the bonds and lone pairs are (a Lewis structure). Sometimes, you are not given a structure to start with. Rather, you just get the molecular formula (for example, $KMnO_4$). The method we saw in the previous section will only work if you can convert the molecular formula into a Lewis structure, so that you can count the electrons. In the next chapter, we will see how to draw a Lewis structure when you are given a molecular formula. But for now, there is a simpler way to calculate oxidation states when you are only given a molecular formula. Rather than drawing the Lewis structure, you can apply some simple rules of thumb.

We will see many rules. In each case, we will focus on why the rule is *a logical consequence* of the method that we used in the previous section. That way, you won't have to memorize the rules. Instead, they will just make sense to you.

We will go over the rules in a methodical order, based on the periodic table. We will start on the right side of the periodic table, and we will work our way slowly to the left (except for the first rule, which applies to all elements). Let's first summarize the rules, and then we will go through them one by one:

- Any neutral, free element will have an oxidation state of zero. Free elements are elements that do not have any bonds to a different type of element. Examples include Mg (when it is a neutral atom, not bonded to any other elements), O_2, S_8, Cl_2, etc. (in each of these cases, there is only one type of element present).

- Fluorine, when present in a compound, will always have an oxidation state of -1. The other halogens will generally have an oxidation state of -1, except when they are connected to F or O.

- Oxygen, when present in a compound, almost always has an oxidation state of -2, except when connected to F. When oxygen is connected to another oxygen atom (like in hydrogen peroxide: HO-OH), the oxidation state is -1.

- Hydrogen, when present in a compound, generally has an oxidation state of $+1$, unless it is connected to a metal or to boron, in which case it will have an oxidation state of -1.

- Elements of Group 2A, when present in a compound, will always have an oxidation state of $+2$.

- Elements of Group 1A, when present in a compound, will always have an oxidation state of $+1$.

Now that we have seen many rules, we need to make sense out of them so that we won't have to memorize any of them.

Let's start with the first rule: *a free element will have an oxidation state of 0*. We can understand why a neutral Mg atom (not bonded to any other elements) will have an oxidation state of zero. After all, it is not sharing any of its electrons, and it is not missing any electrons. So, it will have the number of electrons that it is supposed to have, and the oxidation state will be zero.

But what about Cl_2? Why should the oxidation state for each Cl atom be zero? Let's go back to our method of calculating oxidation states—we treat all bonds as being ionic, except for any bonds between the same type of element. So, a Cl—Cl bond will be split equally:

$$:\overset{..}{\underset{..}{Cl}}-\overset{..}{\underset{..}{Cl}}: \quad \xrightarrow[\text{equally}]{\text{Split this bond}} \quad \boxed{:\overset{..}{\underset{..}{Cl}}\cdot} \ \boxed{\cdot\overset{..}{\underset{..}{Cl}}:}$$

7 electrons

And we see that each chlorine atom will have the number of electrons that it is supposed to have. So the oxidation state of each chlorine atom will be zero.

The first rule tells us that an element can only have a non-zero oxidation state if it has a bond to another kind of element or if it has a charge. After everything we saw in the previous section, this rule is just stating something that should be obvious.

The second rule deals with elements in the second-to-last column of the periodic table. These elements are referred to as the halogens:

Halogens

H																	He
Li	Be											B	C	N	O	F	Ne
Na	Mg											Al	Si	P	S	Cl	Ar
K	Ca	Sc	Ti	V	Cr	Mn	Fe	Co	Ni	Cu	Zn	Ga	Ge	As	Se	Br	Kr
Rb	Sr	Y	Zr	Nb	Mo	Tc	Ru	Rh	Pd	Ag	Cd	In	Sn	Sb	Te	I	Xe
Cs	Ba	La	Hf	Ta	W	Re	Os	Ir	Pt	Au	Hg	Tl	Pb	Bi	Po	At	Rn

The halogens are found in column 7A of the periodic table. This labeling system (using "7A" to label the column) has been used for a long time, and many high school textbooks continue to use this system, so you are probably familiar with it. But since other labeling systems exist, the international community has accepted a standard method for labeling the columns, and it is becoming more common in textbooks (as it should be). This newer system (the IUPAC system) labels the columns from left to right with numbers. With this system, halogens are in the column designated by the number 17. It might be worth one minute of your time to open the inside cover of your textbook and look at the labels on the top of each column of the periodic table. In all of the discussions in *this* book, we will continue to use the old system (but you should be aware of both ways to label columns of the periodic table).

Now, let's get back to the topic at hand: the rule for determining oxidation states of halogens. Let's review the rule: *Fluorine, when present in a compound, always has an oxidation state of −1. The other halogens generally have an oxidation state of −1, unless they are attached to F or O.*

Now, instead of memorizing this rule, let's see why it makes sense. Fluorine has seven valence electrons, so it only needs to form one bond to satisfy its octet. It can form one bond with any other element:

$$X-\ddot{\underset{\cdot\cdot}{F}}{:}$$

Whatever X might be, F will always be more electronegative than X, because F is the most electronegative element. So, when we treat the bond as an ionic bond, we give both electrons of the bond to F. When we do this, we will count eight electrons around F. It is supposed to have only seven electrons, so it will have an oxidation state of −1.

The same logic can be applied to the other halogens. The rest of the halogens are usually −1 as well, because the halogens are more electronegative than most elements (because they are on the right side of the periodic table). This is true EXCEPT when they are connected to F or O, because the halogens are not more electronegative than F and O. So, when connected to F or O, the halogens will have an oxidation state of +1:

$$:\ddot{\underset{\cdot\cdot}{F}}-\ddot{\underset{\cdot\cdot}{Cl}}{:} \xrightarrow[\text{an \textit{ionic bond}}]{\text{Treat this bond like}} :\ddot{\underset{\cdot\cdot}{F}}{:} \quad \boxed{\ddot{\underset{\cdot\cdot}{Cl}}{:}}$$

only six electrons

The oxidation states of halogen can (rarely) get very tricky, because it is possible for many of the halogens (except F) to violate the octet rule and form many bonds. In these situations, the oxidation state will depend on the number and type of bonds. We will discuss this in more detail in Chapter 7 when we talk about the octet rule and violations of it.

For our next rule, we move one column to the left on the periodic table, and we will focus on oxygen. Let's review the rule again: *Oxygen, when present in a compound, almost always has an oxidation state of −2, except when connected to F. When connected to another oxygen atom (like in hydrogen peroxide: HO-OH), it is −1.* Now let's try to make sense of this rule as well.

Oxygen has six valence electrons. So, in order to fill its octet, it will want to form two bonds. It can do this by forming either two single bonds or one double bond:

$$X-\ddot{\underset{\cdot\cdot}{O}}-X \qquad X=\ddot{\underset{\cdot\cdot}{O}}$$

Since oxygen is the second most electronegative element, it will be more electronegative than X, and so, we will end up counting oxygen with eight electrons:

$$X-\ddot{\underset{\cdot\cdot}{O}}-X \xrightarrow[\text{like \textit{ionic bonds}}]{\text{Treat both bonds}} X \quad \boxed{:\ddot{\underset{\cdot\cdot}{O}}{:}} \quad X$$

$$X=\ddot{\underset{\cdot\cdot}{O}} \xrightarrow[\text{like \textit{ionic bonds}}]{\text{Treat both bonds}} X \quad \boxed{:\ddot{\underset{\cdot\cdot}{O}}}$$

In either case, oxygen will have an oxidation state of -2. So, we see that oxygen should always have an oxidation state of -2. The only exception to this would be if X is F or O. When oxygen is connected to another oxygen atom, the two electrons of that bond will be split equally (one on each atom). When we count in these cases, we will find that oxygen will have seven electrons:

So, in compounds that have an O—O bond, oxygen will have an oxidation state of -1. These compounds (an O—O bond) are called peroxides.

We said there is one other exception to oxygen having an oxidation state of -2. That is when oxygen is connected to F. In these cases, the oxidation state of the oxygen will depend on what else the oxygen is connected to (is it another F, an O, or any other element?). You will probably never see any of these examples in your course because O—F bonds are very rare.

Now we can move on to the next rule: *Hydrogen, when present in a compound, has an oxidation state of +1, unless it is connected to a metal or to boron, in which case it will have an oxidation state of −1*. To make sense of this, recall what we said about the electronegativity of H. We said that it is in between B and C. Let's look at the periodic table, and let's place it between the B and the C:

													H				He
Li	Be												B	C	N	O	F	Ne
Na	Mg												Al	Si	P	S	Cl	Ar
K	Ca	Sc	Ti	V	Cr	Mn	Fe	Co	Ni	Cu	Zn	Ga	Ge	As	Se	Br	Kr
Rb	Sr	Y	Zr	Nb	Mo	Tc	Ru	Rh	Pd	Ag	Cd	In	Sn	Sb	Te	I	Xe
Cs	Ba	La	Hf	Ta	W	Re	Os	Ir	Pt	Au	Hg	Tl	Pb	Bi	Po	At	Rn

Hydrogen only forms one bond. When H forms a bond with something to the right (as in the chart above), then it will have an oxidation state of $+1$. However, when it bonds with an element to the left (B or any other metal), then H will be more electronegative. When we try to assign an oxidation state to H, we will find that hydrogen will get both electrons of the bond. H is supposed to have only one electron, so when it has both electrons, it will have an oxidation state of -1.

Hopefully, you are beginning to appreciate that these crazy rules are really not so crazy after all. All of the rules are logical consequences of the simple approach that we saw in the previous section: treating all bonds as if they are ionic, counting up the electrons, and then comparing that number to the number of electrons each atom is supposed to have.

Now let's move on to our next rule: *Elements of Group 2A, when present in a compound, will always have an oxidation state of +2*. To make sense of this, let's look at the periodic table again.

2A

H																	He
Li	Be											B	C	N	O	F	Ne
Na	Mg											Al	Si	P	S	Cl	Ar
K	Ca	Sc	Ti	V	Cr	Mn	Fe	Co	Ni	Cu	Zn	Ga	Ge	As	Se	Br	Kr
Rb	Sr	Y	Zr	Nb	Mo	Tc	Ru	Rh	Pd	Ag	Cd	In	Sn	Sb	Te	I	Xe
Cs	Ba	La	Hf	Ta	W	Re	Os	Ir	Pt	Au	Hg	Tl	Pb	Bi	Po	At	Rn

These elements are called the alkaline earth metals, and they always form two bonds (they have two valence electrons, and they use both of them to form bonds). It is hard to imagine one of these elements sharing a bond with an element in the first column of the periodic table. You will never see an example of that. So, they will always be bonded to elements to the right on the periodic table. Therefore, when we try to assign an oxidation state, we will see that they lose both electrons, to get an oxidation state of +2:

$$:\ddot{\text{Cl}}-\text{Mg}-\ddot{\text{Cl}}: \xrightarrow[\textit{ionic bonds}]{\text{Treat these bonds like}} :\ddot{\ddot{\text{Cl}}}: \boxed{\text{Mg}} :\ddot{\ddot{\text{Cl}}}:$$

no electrons

Be careful though. If one of the alkaline earth metals is not part of a compound, then it will have an oxidation state of 0. Mg, by itself, will have an oxidation state of 0 (that was the first rule we saw). It will only have an oxidation state of +2 when it is sharing bonds with other atoms.

And now we get to our last rule: *Elements of Group 1A, when present in a compound, will always have an oxidation state of +1*.

1A

																	He
Li	Be											B	C	N	O	F	Ne
Na	Mg											Al	Si	P	S	Cl	Ar
K	Ca	Sc	Ti	V	Cr	Mn	Fe	Co	Ni	Cu	Zn	Ga	Ge	As	Se	Br	Kr
Rb	Sr	Y	Zr	Nb	Mo	Tc	Ru	Rh	Pd	Ag	Cd	In	Sn	Sb	Te	I	Xe
Cs	Ba	La	Hf	Ta	W	Re	Os	Ir	Pt	Au	Hg	Tl	Pb	Bi	Po	At	Rn

These elements are called the alkali metals (not to be confused with the alkali**ne earth** metals in column 2A). The logic we use here is similar to the logic we used above. When an element from column 1A is part of a compound (in other words, when it shares a bond with another element), it will always have an oxidation state of +1.

Now we have seen why the rules make sense. Remember that the whole point of these rules was to have a quick way to assign oxidation states when we are only given a molecular formula (but no Lewis structure). To see how we use the rules to help in those situations, we need one last rule: *For any molecule or ion, the sum total of all of the oxidation states must equal the charge on the molecule or ion.* For neutral molecules (no charge), this rule tells us that the sum of the oxidation states must be zero. Let's focus on neutral molecules first, and then we will focus on ions later. Let's see an example of how we use all of these rules:

EXERCISE 6.38. Calculate the oxidation state of Mn in potassium permanganate ($KMnO_4$):

Answer: Overall, the compound does not have a net charge (it would be shown if there was), so the sum of all of the oxidation states must be zero. Using our rules, we can say that the oxidation state of K must be +1 (it is an alkali metal), and the oxidation state of each oxygen atom is probably −2. Assume that the oxygen atoms are not bonded to each other (you can usually make this assumption). So, the sum total of the oxidation states (excluding Mn) will be $(+1) + 4(-2) = -7$. In order for the sum total of the oxidation states to be 0, the Mn atom must have an oxidation state of +7.

Notice that you are able to use the rules of thumb to calculate the oxidation state for an atom in a molecule, *without* counting the electrons in the Lewis structure of the molecule.

PROBLEM 6.39. Calculate the oxidation state of Cr in CrO_3.

PROBLEM 6.40. Calculate the oxidation state of S in H_2SO_4.

PROBLEM 6.41. Calculate the oxidation state of Zn in $ZnCl_2$.

PROBLEM 6.42. Calculate the oxidation state of B in BF_3.

PROBLEM 6.43. Calculate the oxidation state of N in HNO_3.

PROBLEM 6.44. Calculate the oxidation state of C in $COCl_2$.

In order to do the problems above, we said that the sum total of all the oxidation states in a neutral molecule must be zero. Now, let's consider what to do in cases where we have an ion rather than a neutral molecule. We just said that the sum total of all of the oxidation states must equal the charge on the ion. As an example, consider the ion CO_3^{2-}. The sum total of all of the oxidation states must be -2, because that is the net charge on the ion (as indicated). Now, let's see an example of this:

EXERCISE 6.45.   Calculate the oxidation state of C in the carbonate ion (CO_3^{2-}).

Answer: This ion has a net charge of -2, so the sum of all of the oxidation states must equal -2. Other than the carbon atom, there are only three oxygen atoms. Once again, you can assume that they are not connected to each other (each of them is actually connected to the central carbon atom), so each oxygen atom will have an oxidation state of -2.

So, the sum total of the oxidation states (excluding carbon) will be $3(-2) = -6$. In order for the sum total of the oxidation states to be -2 (which is the charge of the ion), the carbon atom must have an oxidation state of $+4$.

PROBLEM 6.46.   Calculate the oxidation state of the nitrogen atom in NO_3^-.

PROBLEM 6.47.   Calculate the oxidation state of the sulfur atom in SO_4^{2-}.

PROBLEM 6.48.   Calculate the oxidation state of the aluminum atom in $AlCl_4^-$.

PROBLEM 6.49.   Calculate the oxidation state of the boron atom in BH_4^-.

PROBLEM 6.50.   Calculate the oxidation state of the nitrogen atom in NO_2^+.

6.7   AN INTRODUCTION TO RED-OX REACTIONS

In the previous section, we learned the rules for assigning oxidation states. This skill will be important when we learn about reactions, because it will allow us to keep track of the electrons during a reaction. It makes sense that we should want to focus on the electrons during a reaction. After all, a reaction is when bonds are broken and/or formed (and bonds are just overlapping electrons). So any reaction can be described in terms of the movement of the electron density. Clearly, in order to understand a reaction, we will want to know how the electron density has moved during the reaction. By looking at oxidation states before **and** after a reaction takes place, we get a better understanding of how the electron density has moved.

In some reactions, we will find that elements change their oxidation states throughout the course of the reaction. These reactions are called oxidation-reduction reactions. For example, consider the following reaction:

$$Mg + Cl_2 \longrightarrow MgCl_2$$

Let's first consider the oxidation state of the Mg atom. Before the reaction takes place, the Mg atom is a neutral, free element. Therefore, it has an oxidation state of zero. After the reaction, the Mg atom is bonded to two electronegative chlorine atoms. So, the oxidation state of the Mg atom in $MgCl_2$ will be $+2$.

We see that the oxidation state of the Mg atom went from 0 to $+2$. An increase in oxidation state is called an ***oxidation***. So, we would say that the Mg atom was *oxidized* in this reaction.

Now let's consider what happens to the chlorine atoms. Before the reaction, each chlorine atom in Cl_2 is considered a free element, and will have an oxidation state of zero. After the reaction, each chlorine atom will have an oxidation state of -1. So, the chlorine atoms have gone from an oxidation state of 0 an oxidation state of -1. A reduction in oxidation state is called a **reduction**.

Notice that the Mg atom was *oxidized*, and the chlorine atoms were *reduced*. In fact, oxidation is *always* accompanied by reduction. In order for something to get oxidized, it must happen at the expense of something else (which will be reduced). Later in this course, you will see many reduction/oxidation reactions (often called **redox** reactions, which is short for **red**uction-**ox**idation). In order to understand those reactions, you will need to be able to look at a reaction and locate which elements are oxidized and which are reduced during the reaction. Now we need to make sure that we have that skill. Let's see an example:

EXERCISE 6.51. In the reaction below, identify which element is reduced and which element is oxidized:

$$CH_4 \ (g) + 2O_2 \ (g) \longrightarrow CO_2 \ (g) + 2H_2O \ (g)$$

Answer: In order to do this, you will need to calculate the oxidation state of every atom, and you will need to look for the changes. If you calculate every oxidation state (using the rules and skills that you have developed earlier), you will get:

$$\overset{-4\ +1}{CH_4} \ (g) + \overset{0}{2O_2} \ (g) \longrightarrow \overset{+4\ -2}{CO_2} \ (g) + \overset{+1\ -2}{2H_2O} \ (g)$$

Notice that we are referring to the oxidation state of each individual atom. So, you will notice above that the H in H_2O has a $+1$ written over it. This means that each hydrogen atom in the molecule has an oxidation state of $+1$. When we are looking at the oxidation states of each atom, the coefficients (that balance the reaction) are not important. So, the fact that H_2O has a coefficient of 2 does not change anything. For now, you can just ignore that 2.

Now that you have calculated the oxidation state of each atom in the reaction, you are ready to take a closer look at each of the elements present in this reaction. There are three elements: C, H, and O. Let's look at each of them individually.

The carbon atom went from an oxidation state of -4 to $+4$. Since it increased its oxidation state, it was *oxidized*.

The hydrogen atoms went from an oxidation state of $+1$ to an oxidation state of $+1$. So they were not oxidized or reduced.

The oxygen atoms went from an oxidation state of zero to an oxidation state of -2. Since their oxidation states were reduced, we can say that the oxygen atoms were *reduced*.

PROBLEM 6.52. In the reaction below, identify which element is reduced and which element is oxidized:

$$2HNO_3 \ (aq) + 3H_2S \ (g) \longrightarrow 2NO \ (g) + 4H_2O \ (l) + 3S \ (s)$$

Remember that the coefficients are irrelevant for the calculations that you are doing here, but the subscripts in the molecular formula are very important. As an example: in the first compound, you can ignore the coefficient of 2, but you must consider the subscript of 3 next to the oxygen when you are calculating the oxidation of the nitrogen atom in that compound.

Answer: _____

PROBLEM 6.53.   In the reaction below, identify which element is reduced and which element is oxidized:

Answer: _____

PROBLEM 6.54.   In the reaction below, identify which element is reduced and which element is oxidized:

$$K_2Cr_2O_7 + 14HCl \longrightarrow 2CrCl_3 + 3Cl_2 + 7H_2O + 2KCl$$

Answer: _____

PROBLEM 6.55.   In the reaction below, identify which element is reduced and which element is oxidized:

$$NaNO_3 + 4Zn + 7NaOH \longrightarrow NH_3 + 4Na_2ZnO_2 + 2H_2O$$

Answer: _____

PROBLEM 6.56.   In the reaction below, identify which element is reduced and which element is oxidized:

Answer: _____

CHAPTER 7

DRAWING LEWIS STRUCTURES

For many types of problems in this course, you will need to convert a molecular formula into a Lewis structure:

C_2H_5OH

Molecular Formula   *Lewis Structure*

The Lewis structure of a compound shows useful information: it tells us which atoms are connected together, and more importantly, where the electrons are. In this section, we will focus on the skills that you need to draw Lewis structures (frequently called Lewis diagrams or Lewis formulas).

7.1   A DEEPER UNDERSTANDING OF THE OCTET RULE

You probably remember learning about the octet rule in high school. You probably also remember that there are exceptions to the rule. The truth is that there are MANY exceptions to the rule. This should prompt us to ask several questions: Why do we have a rule that gets broken so often? *When* does it get broken, and when is it followed? And what exactly is this rule in the first place? Where did the rule come from?

In this section, we will take a closer look at the octet rule. We will try to understand it better, so that we can predict *how*, *why*, and *when* it can be broken. This will be critical for drawing Lewis structures. But first, we need to quickly review some information that we saw in the previous chapter.

We saw that electrons exist in predefined regions of space, called *orbitals*. Every orbital can hold a maximum of two electrons. So, any orbital can have one of the following possibilities: the orbital can contain (a) no electrons, (b) one electron, or (c) two electrons. These are the only possibilities, since an orbital can never have more than two electrons. Let's quickly consider each of these situations, because they will prove relevant for drawing and appreciating Lewis structures.

• If an orbital contains *no electrons*, then there is nothing to talk about. Remember that an orbital is just a predefined region of space where 95% of the *electron cloud* exists (you can think of the orbital like a house, but *without* walls or a roof). If the electron is not home, then there is nothing there. No walls, no roof. Nothing.

- If an orbital has **one electron**, then it is ready to form a bond. Remember how bonds are formed. We use an atomic orbital containing one electron to overlap with an atomic orbital from another atom (which also contains one electron). The overlap creates two new orbitals called molecular orbitals (we used the analogy of a mansion and an out-house), and both electrons go into the bonding molecular orbital.

- If an orbital has **two electrons**, then we call it a *lone pair*, and the electrons are non-bonding. Each lone pair is shown as a pair of dots:

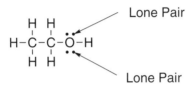

To summarize, we have seen that any orbital can have no electrons, one electron (which will be used for bonding), or two electrons (which will remain as a non-bonding lone pair). We will soon see that the octet rule (and its violations) will be more understandable by focusing on the *orbitals*. Let's look closely at the orbitals for the *second-row* elements:

H																	He
Li	Be											B	C	N	O	F	Ne
Na	Mg											Al	Si	P	S	Cl	Ar
K	Ca	Sc	Ti	V	Cr	Mn	Fe	Co	Ni	Cu	Zn	Ga	Ge	As	Se	Br	Kr
Rb	Sr	Y	Zr	Nb	Mo	Tc	Ru	Rh	Pd	Ag	Cd	In	Sn	Sb	Te	I	Xe
Cs	Ba	La	Hf	Ta	W	Re	Os	Ir	Pt	Au	Hg	Tl	Pb	Bi	Po	At	Rn

The valence electrons of these elements can exist in **four** orbitals. Why four? The *second row* elements use orbitals in the *second energy level*, and that level has one s orbital and three p orbitals, **for a total of four orbitals**. We know that the periodic table is organized so that as we move across a row, each element gets one more valence electron than the previous element. We use dots to show the valence electrons, and starting with Li, we place one electron per orbital until we get to carbon:

For each of the elements above, it would seem as though we are *not* drawing one s orbital and three p orbitals. The drawings above are *not* meant to illustrate the actual shapes of the orbitals. That would require a more detailed discussion, and we will discuss this in Chapter 8, which deals with hybridization states. For now, let's ignore the shapes of the orbitals, and let's just focus on the *number* of electrons in each orbital. When we look at the drawings above, we can see that carbon (unlike Li, Be, and B) has no empty orbitals. Each of the four orbitals of carbon is occupied by one electron.

As we continue along the second row, we begin to pair up electrons (as lone pairs) until we get to the noble gas configuration of Ne:

When we inspect all of the drawings above, we can clearly see which orbitals are empty, which orbitals have one electron, and which orbitals have two electrons. For example, look at beryllium above. It has two empty orbitals and two orbitals that each have one electron. Now look at nitrogen. It has no empty orbitals. All four orbitals are being used. One orbital has a lone pair, and the other three orbitals have one electron each, ready to form bonds.

When we draw Lewis structures, we generally don't draw the orbitals the way we did above. We just draw the dots to show the electrons:

Li·   Be·   ·B·   ·Ċ·   ·N̈·   ·Ö·   :F̈·   :N̈e:

Believe it or not, chemists pay so much attention to the octet rule because of the behavior of *just a few elements* on the periodic table. Just a few of the elements above will always follow the octet rule. Obviously, we need to take a closer look at why this is the case.

You probably remember the definition of the octet rule from your chemistry class in high school. The most stable arrangement for an atom is when it has the configuration of a noble gas (a filled valence shell, like neon above). But why? Why should the atom care?

To understand this we need to revisit the analogy we gave in Chapter 6 (remember the tornado that formed a mansion and an out-house?). We saw that when an electron exists by itself in an atomic orbital, it can lower its energy by forming a bond. We used the analogy of a tornado turning two houses into a mansion and out-house, as a way to explain how atomic orbitals overlap to form molecular orbitals. And we saw that the bonding molecular orbital is lower in energy than the original two atomic orbitals:

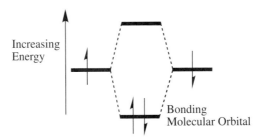

Increasing Energy

Bonding
Molecular Orbital

As a side point, you might be wondering *why* an electron cares whether or not it is in a lower energy state? We always talk about electrons or molecules preferring lower energy states—but why? The answer to this question is right around the corner (you will have the answer to this question when you learn about entropy and the second law of thermodynamics). For now, we just need to accept this concept as a given.

So we see that a single electron in an atomic orbital will not prefer to remain alone as an unpaired electron. Rather, it will form a bond to lower its energy. We have also seen that a lone pair will generally not bond with another atom. It will remain as a lone pair. These two concepts are the foundational concepts to reframe the octet rule in terms of the *orbitals*. As an example, let's consider the nitrogen atom again:

Nitrogen has five valence electrons that get distributed over four orbitals. We said that one orbital has a lone pair (circled above to emphasize it), and the other three orbitals have one electron each. The lone pair will generally be happy the way it is, and it will not need to form a bond. But the other three orbitals (that each contain one electron) will certainly want to form bonds. And that is why we see so many structures where nitrogen has one lone pair and three bonds:

$$H \cdot \quad \cdot \overset{\cdot\cdot}{\underset{\cdot}{N}} \cdot \quad \cdot H \longrightarrow H - \overset{\cdot\cdot}{\underset{|}{N}} - H$$
$$\overset{\cdot}{H} \qquad\qquad\qquad H$$

A similar argument can be made for oxygen. Oxygen has six valence electrons:

We see that two orbitals are filled with lone pairs, and the other two orbitals have one electron each. These two orbitals will want to form bonds, and that explains why we see examples of oxygen with two bonds and two lone pairs:

$$\text{H}\cdot\ \ \ddot{\ddot{\text{O}}}\cdot\ \cdot\text{H} \quad \longrightarrow \quad \text{H}-\overset{\cdot\cdot}{\underset{\cdot\cdot}{\text{O}}}-\text{H}$$

You can also make a similar argument for fluorine. It has three lone pairs, and one orbital containing one electron. So, fluorine will form only one bond:

$$\ddot{\text{F}}\cdot\ \ \cdot\text{H} \quad \longrightarrow \quad \ddot{\text{:F}}-\text{H}$$

In every one of these cases, we see that the end product is that each of these elements gets surrounded by eight electrons:

$$\text{H}-\overset{\cdot\cdot}{\underset{|}{\text{N}}}-\text{H} \qquad \text{H}-\overset{\cdot\cdot}{\underset{\cdot\cdot}{\text{O}}}-\text{H} \qquad \ddot{\text{:F}}-\text{H}$$
$$\quad\ \ \text{H}$$

In all of these cases, we have seen one underlying theme: ***all four orbitals want to be filled with two electrons*** one way or another; whether they are filled as lone pairs (non-bonding), or whether they are filled with two shared electrons from a bond. And that is why these elements want to have an octet.

By reframing the octet rule in terms of the orbitals, we can see that all of the exceptions to the octet rule will make sense. First of all, we can now understand why second-row elements can NEVER have five bonds. Remember that a bond can only be formed from an orbital. A second-row element has only four orbitals to use, so a second-row element will ***never*** form *more* than four bonds. And there are NO EXCEPTIONS to this.

It ***is*** possible for a second-row element to have *less* than eight electrons. In fact, the first three elements in the second row *regularly* fail to fill all four orbitals. Let's start with the first element: Li. Lithium has one valence electron because it is in the first column of the periodic table:

H																	He
Li	Be											B	C	N	O	F	Ne
Na	Mg											Al	Si	P	S	Cl	Ar
K	Ca	Sc	Ti	V	Cr	Mn	Fe	Co	Ni	Cu	Zn	Ga	Ge	As	Se	Br	Kr
Rb	Sr	Y	Zr	Nb	Mo	Tc	Ru	Rh	Pd	Ag	Cd	In	Sn	Sb	Te	I	Xe
Cs	Ba	La	Hf	Ta	W	Re	Os	Ir	Pt	Au	Hg	Tl	Pb	Bi	Po	At	Rn

There is no way that Li is ever going to form four bonds. It might have four orbitals available for bonding, but it just doesn't have the electrons to form those bonds. It would need three more electrons to be able to form a total of four bonds. And if we gave it three more electrons, it would have a charge of -3. That just isn't going to happen. So, lithium will NEVER have eight electrons around it.

The same argument can be applied to the next element in the second row. Beryllium has only two valence electrons, because it is in the second column (2A) of the periodic table. In order to form four bonds, we would need to give it two more electrons. If we did that, it would have a charge of -2. That is actually possible, but very hard to do. So you just won't see common examples of Be with four bonds. And therefore, Be will generally not fill its octet either.

Now we move along to the next element in the second row, boron:

H																	He
Li	Be											B	C	N	O	F	Ne
Na	Mg											Al	Si	P	S	Cl	Ar
K	Ca	Sc	Ti	V	Cr	Mn	Fe	Co	Ni	Cu	Zn	Ga	Ge	As	Se	Br	Kr
Rb	Sr	Y	Zr	Nb	Mo	Tc	Ru	Rh	Pd	Ag	Cd	In	Sn	Sb	Te	I	Xe
Cs	Ba	La	Hf	Ta	W	Re	Os	Ir	Pt	Au	Hg	Tl	Pb	Bi	Po	At	Rn

Boron is a different situation, because it has *three* valence electrons. So it will comfortably form *three* bonds:

$$\underset{H}{\overset{H}{\underset{\diagdown}{\overset{|}{B}}}}\,H \qquad \underset{F}{\overset{F}{\underset{\diagdown}{\overset{|}{B}}}}\,F$$

In the examples above, boron does *not* have an octet of electrons. And that's OK. Call it a *violation* of the octet rule, if you like. But it makes sense. If we want boron to form four bonds, then we would have to give it one more electron, and then it would have a formal charge of -1. This is more acceptable than the situation with Li or Be, because it is very common for atoms to have a formal charge of -1. In fact, there are many examples where boron has four bonds and a negative charge:

$$H-\overset{\overset{\textstyle H}{|}}{\underset{\underset{\textstyle H}{|}}{B}}^{\ominus}\!H \qquad F-\overset{\overset{\textstyle F}{|}}{\underset{\underset{\textstyle H}{|}}{B}}^{\ominus}\!F$$

And in these cases boron is said to *obey* the octet rule, because it is surrounded by eight electrons. So, we see that boron will *sometimes* obey the octet rule (when it has a formal charge of -1). We will see more on this later in this chapter.

For now, let's move on to the next element in the second row (after boron), which is carbon. Since carbon has one more electron than boron, we now have *four* valence electrons to work with. So with carbon, we will have no problem using all four orbitals. Carbon will always form four bonds, and it will *not* have a formal charge the way that boron did. Here is a simple example:

$$
\begin{array}{c}
\text{H} \\
| \\
\text{H}-\text{C}-\text{H} \\
| \\
\text{H}
\end{array}
$$

There are cases when carbon can have *less* than eight electrons around it, but these are very reactive intermediates. They don't survive for very long before they react to form an octet again. You will learn about these reactive intermediates next year in organic chemistry.

We have already discussed the situations for nitrogen, oxygen and fluorine, and we saw that they always have an octet. They will not have four bonds like carbon, because they have lone pairs (which generally don't form bonds). But, they still have all four orbitals being utilized, with two electrons in each orbital:

$$
\text{H}-\overset{..}{\underset{|}{\text{N}}}-\text{H} \qquad \text{H}-\overset{..}{\underset{..}{\text{O}}}-\text{H} \qquad \overset{..}{\underset{..}{:\text{F}}}-\text{H}
$$
$$
\text{H}
$$

Let's summarize what we have seen so far:

- Li and Be will **never** fill their octet.
- B will **sometimes** fill its octet, but only when it has a negative charge.
- C, N, O, and F will **always** fill their octet

So, we see that the entire hoopla over the octet rule is really just over a few elements: (B), C, N, O, and F. Once you move into the third row of the periodic table, you have access to more orbitals (the third energy level has five d orbitals, in addition to one s orbital and three p orbitals). Those elements are able to form *more* than four bonds.

So, the obvious question is: why do we have a rule that only applies to four elements in all situations? The periodic table has so many elements, that we should anticipate that there will be more exceptions to the rule than examples of the rule being followed! The answer is simple. C, N, O, and F are the most common elements, and most of the compounds that you will encounter will use one or more of these elements. Therefore it is important to understand their nature. They only have four orbitals to use, so they will NEVER form five bonds, because that would require a fifth orbital. Similarly, these elements will NEVER form four bonds and one lone pair. That would also require a fifth orbital, which does not exist. As well, these elements will NEVER form three bonds and two lone pairs. Once again, there is no fifth orbital.

When we talk about *violations* to the octet rule, we have to understand what we are talking about. Suppose we were to define the octet rule like this: ***second-row elements will never exceed an octet, because they only have four orbitals to utilize***. This definition is NEVER violated. If we had expressed the octet rule in this way, then we would never have any violations to this rule. But, chemists don't express the rule like this. Instead, we say that elements will prefer to have an octet. If you say the rule like that, then there will be MANY violations. For starters, we saw that Li and Be will always violate this rule. Boron will sometimes violate this rule. And the third-row elements have access to *more* than four orbitals, so they can have *more* than eight electrons at certain times, which is yet another violation.

When we say that there are *violations* to the octet rule, this is really a semantic argument based on the way we chose to express the idea that second-row elements only have four orbitals to use.

Now we can appreciate that there are two kinds of *violations* to the rule as chemists have defined it. First of all, we have elements that can have *less* than an octet of electrons (like Li, Be, and sometimes B). Then we have elements in the third row (like P, S, and Cl) that can regularly have *more* than an octet of electrons. It is important that we learn how to spot these elements. Whenever you see the former group of elements, you should think to yourself that it's OK if they don't have a full octet. Whenever you see the latter group of elements, you should think to yourself that it's OK if they have *more* than an octet. We need to get practice recognizing these elements. To help out, carefully study the following trends, which should make sense based on everything we have said:

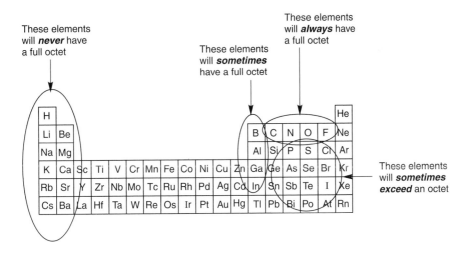

Let's get some practice recognizing these elements. We will now look at a long list of compounds, and we will try to identify the elements that fit into the categories shown in the chart above (1- never has an octet, 2- sometimes has an octet, 3- always has an octet, and 4- can sometimes exceed an octet). Obviously, this will be a trivial task if you constantly look up each element on the periodic table. That would defeat the purpose of this exercise. Instead, study the periodic table trends above for a little while. Make sure you know which parts of the periodic table correspond to each category. Then try to identify the elements in the compounds given below *without* looking at the chart above. Instead, use a different periodic table (either from your textbook or in the front cover of this book). For each element in every compound below, use that periodic table to locate the element, and then try to make a decision regarding what category that element is in.

There are many compounds below. You will be at a significant advantage in the upcoming chapters if you can identify elements and immediately be able to tell whether they will follow the octet rule or not. You do not need to go through *all* of the exercises below. You can stop whenever you feel that you have gained a familiarity with the positions of the elements on the periodic table.

PROBLEM 7.1.   For each element in the compounds below, determine which category best describes the nature of the element. The four categories are:

   1. It *never* has an octet.

   2. It *sometimes* has an octet.

   3. It *always* has an octet.

   4. It can sometimes *exceed* an octet.

a.	SO_2F_2	**n.**	NO_2F	**aa.**	CH_4S
b.	COS	**o.**	H_2SO_4	**ab.**	NO_2Cl
c.	$AlBr_3$	**p.**	H_3SeO_4	**ac.**	IF_4
d.	$HAsO_4$	**q.**	BeF_2	**ad.**	H_3PO_3
e.	$HClO_3$	**r.**	H_2SeO_3	**ae.**	$HBrO_4$
f.	H_2CO_3	**s.**	HNO_3	**af.**	H_3PO_4
g.	$CO_3{}^{2-}$	**t.**	$NO_3{}^-$	**ag.**	$H_2PO_4{}^-$
h.	$HCO_3{}^-$	**u.**	$IO_3{}^-$	**ah.**	$HSO_4{}^-$
i.	H_2CO_3	**v.**	HNO_2	**ai.**	$COCl_2$
j.	$NH_4{}^+$	**w.**	$AlH_4{}^-$	**aj.**	$PCl_4{}^+$
k.	CS_2	**x.**	SiF_4	**ak.**	$BF_4{}^-$
l.	$NO_2{}^-$	**y.**	$ClO_3{}^-$	**al.**	$CO_3{}^{2-}$
m.	$ClO_2{}^-$	**z.**	$PO_4{}^{3-}$	**am.**	$PO_3{}^{3-}$

Now we will learn a method for drawing Lewis structures. To make things easier at first, we will learn the method and practice on compounds that do *not* contain elements that violate the octet rule. Once we have learned the method for drawing Lewis structures, then we will revisit the elements that violate the octet rule.

7.2 LEWIS DOT DIAGRAMS OF BINARY COMPOUNDS

Let's start out with very small molecules that consist of just two atoms bonded to-gether. For example, F_2 is a molecule consisting of just two atoms. So is N_2. But if we compare the Lewis structures of these two compounds, we will see that they are fairly different:

$$:\overset{\cdot\cdot}{\underset{\cdot\cdot}{F}}-\overset{\cdot\cdot}{\underset{\cdot\cdot}{F}}: \qquad :N\equiv N:$$

Notice that F_2 has only a single bond and many lone pairs, but N_2 has a triple bond and only a couple of lone pairs. How are you supposed to know this without see-ing the structures first? There is a simple method for determining whether to place electrons as bonds or as lone pairs. And it goes like this:

Begin by counting the number of valence electrons that each atom has. We saw how to do this in Chapter 6 when we were calculating oxidation states and for-mal charges. We saw that the periodic table tells you how many valence electrons each element is supposed to have. When we look at the periodic table, we see that F has seven valence electrons, but N only has five valence electrons:

H															5A		7A	He
Li	Be												B	C	N	O	F	Ne
Na	Mg												Al	Si	P	S	Cl	Ar
K	Ca	Sc	Ti	V	Cr	Mn	Fe	Co	Ni	Cu	Zn	Ga	Ge	As	Se	Br	Kr	
Rb	Sr	Y	Zr	Nb	Mo	Tc	Ru	Rh	Pd	Ag	Cd	In	Sn	Sb	Te	I	Xe	
Cs	Ba	La	Hf	Ta	W	Re	Os	Ir	Pt	Au	Hg	Tl	Pb	Bi	Po	At	Rn	

Let's start with case of fluorine, and then we will come back to the case of nitrogen. Each fluorine atom comes into the game with seven valence electrons. There are two fluorine atoms in our molecule, so that gives us a total of 14 elec-trons that need to be shown in our drawing. Next, we ask: how many electrons would it take to give every atom an octet (giving each atom eight electrons)? In this case, there are only two atoms, so $2 \times 8 = 16$. To summarize, we have seen that we need to have 16 electrons in order to satisfy every element, but we only have 14 electrons to work with. Therefore, we are missing *two* electrons. That tells us that *two* electrons need to be shared. Remember that every bond is two electrons, so, if we only need to share two electrons, then we will only need one bond. We start by drawing that single bond:

$$F-F$$

Then, we fill in the rest of the electrons as lor̄ ᵖairs, in such a way that gives every atom an octet of electrons. We originally had ⸱ valence electrons, and we used two of these electrons to form a bond. So, we are left with 12 electrons that need to be used as lone pairs. 12 electrons = 6 lone pairs, because every lone pair is two electrons. So, we place three lone pairs on each fluorine atom, and we see that each fluorine atom will have its octet:

$$:\!\ddot{F}\!-\!\ddot{F}\!:$$

And that is why we draw F_2 with only one bond and many lone pairs. As a final check, we will count the number of electrons in our drawing. We see that there is one bond (2 electrons) and 6 lone pairs (12 electrons), which gives a total of 14. And that is how many electrons we started with, so we know we didn't make a mistake when placing the electrons. This last step might seem trivial, but it is very important to get in the habit of always doing this, because the most common mistake is to place the wrong number of electrons on the Lewis structure.

Now let's apply the same logic to N_2. We start by counting the number of valence electrons. Each nitrogen atom comes into the game with *five* valence electrons, which gives a total of 10 electrons tha̅ ̅eed to be shown in our drawing. Next, we ask ourselves: how many electrons would it take to give every atom an octet? In this case, there are only two atoms, so $2 \times 8 = 16$. To summarize, we need to have 16 electrons in order to satisfy every element, but we only have 10. We are missing 6 electrons. That tells us thaᵗ 6 electrons need to be shared. Remember that every bond is 2 electrons, so, if ⸱ need to share 6 electrons, then we will need 3 bonds. So we start by drawing these 3 bonds:

$$N\equiv N$$

Then, we fill in the rest of the electrons as lone pairs, in such a way that gives every atom an octet of electrons. We originally had 10 valence electrons, and we used 6 of these electrons to form bonds. So, we are left with 4 electrons that need to be used as lone pairs. That is 2 lone pairs. When we place one lone pair on each nitrogen atom, we see that each nitrogen atom will have its octet:

$$:\!N\equiv N\!:$$

And that is why we draw N_2 with 3 bonds and only a couple of lone pairs. As a final check, we will count the number of electrons in our drawing. We see that there are 3 bonds (6 electrons) and 2 lone pairs (4 electrons), which gives a total of 10 electrons. And that is how many electrons we started with, so we know we did not make a silly mistake.

Now let's review our method: Start by counting the total number of valence electrons. Then count how many electrons are needed for every atom to have an octet. Based on the difference between these two numbers, you will know how many electrons need to be shared as bonds. Then place the rest as lone pairs so that every atom gets an octet. In the end, don't forget to do a final check to make sure you used the right number of electrons.

Before we do some exercises, there is one more point to make. Hydrogen does not get eight electrons. Hydrogen only uses one orbital (the first energy level has only one s orbital), so it only needs two electrons.

EXERCISE 7.2. Draw the Lewis structure for HCl.

Answer: Start by counting the total number of electrons that you have to work with. H has 1 valence electron, and Cl has 7 valence electrons. So there are a total of 8 electrons that need to be drawn in your structure. Next, calculate how many electrons you need in order to give every atom an octet. Cl needs 8 electrons, and remember that H only needs 2 electrons. So, you need to have 10 electrons in order to satisfy every element, but you only have 8. You are missing 2 electrons. That tells you that 2 electrons need to be shared, which means that you will need only 1 bond. So, start by drawing this bond:

$$H-Cl$$

Then, fill in the rest of the electrons as lone pairs, in such a way that satisfies every atom. You originally had 8 valence electrons, and you used 2 of these electrons to form a bond. So, you are left with 6 electrons that need to be used as lone pairs. That is 3 lone pairs. The hydrogen atom is already happy, because it already has 2 electrons, so place all of the lone pairs on the chlorine atom. This gives chlorine an octet:

$$H-\ddot{\underset{\cdot\cdot}{Cl}}:$$

As a last step, always double-check your answer to make sure you counted the electrons properly. The drawing shows 8 electrons, and that is how many you started with, so you know that you didn't make a mistake along the way.

PROBLEM 7.3. Draw the Lewis structure for Br_2.

PROBLEM 7.4. Draw the Lewis structure for I_2.

PROBLEM 7.5. Draw the Lewis structure for HF.

PROBLEM 7.6. Draw the Lewis structure for BrCl.

PROBLEM 7.7. Draw the Lewis structure for HI.

The case of molecular oxygen (O_2) is a special case, and it doesn't conform so nicely to the method just shown. At some point, you will learn about the paramagnetic behavior of molecular oxygen, and you will see that this can best be understood using molecular orbital theory to explain the structure of oxygen. We will not get into that whole discussion here, but when the time comes, you should review the first few pages of Chapter 6, so that you will be more prepared to read about molecular orbital theory in your textbook.

In all of the previous problems we were dealing with neutral molecules. But sometimes, you will have to draw Lewis structures for ions as well. The method is identical; you just need to take the charge into account when you count the valence electrons. To see how this works, let's see an example:

EXERCISE 7.8.   Draw the Lewis structure for the hydroxide ion (HO^-).

Answer: Start by counting the total number of electrons that you have to work with. H has 1 valence electron, and O has 6 valence electrons. But there is a charge of -1 on this ion, which tells you that there is one extra electron here. So, when you add up the electrons, you get: $1 + 6 + 1$. This means that there is a total of 8 electrons that need to be drawn in your structure.

Next, calculate how many electrons you need in order to give every atom an octet. O needs 8 electrons, and H needs 2 electrons. So, you need to have 10 electrons in order to satisfy every element, but you only have 8. You are missing 2 electrons. That tells you that 2 electrons need to be shared, which means that you will need only one bond. So, start by drawing this bond:

$$H-O$$

Then, fill in the rest of the electrons as lone pairs in such a way that satisfies every atom. You originally had 8 valence electrons, and you used 2 of these to form a bond. So, you are left with 6 electrons that need to be used as lone pairs; 6 electrons = 3 lone pairs. The hydrogen atom is already happy, because it already has 2 electrons, so place all 3 lone pairs on the oxygen atom. This gives oxygen an octet:

$$H-\overset{..}{\underset{..}{O}}:$$

Now you need to show that this ion is negatively charged, so you need to draw the negative charge. Recall from Chapter 6 that you learned how to assign formal charges. If you apply those rules here, you will see that oxygen has a formal charge of -1, and that is why the ion has an overall charge of -1. So, place the negative charge on the oxygen:

$$H-\overset{..}{\underset{..}{O}}:^{\ominus}$$

As a last step, always double-check your answer to make sure you counted electrons properly. The drawing shows 8 electrons, and that is how many you started with, so you know that you didn't make a mistake along the way. (The negative charge at this point does not indicate that you should add an electron—there is a negative charge there because the oxygen atom has one extra electron that is already shown in your drawing.)

PROBLEM 7.9. Draw the Lewis structure for the cyanide ion (CN^-).

PROBLEM 7.10. Draw the Lewis structure for the hypochlorite ion (ClO^-).

PROBLEM 7.11. Draw the Lewis structure for HS^-.

PROBLEM 7.12. Draw the L   re for NO$^+$.

7.3 SELECTING THE CENTRAL ATOM IN COMPOUNDS

In the previous section, we focused on molecules consisting of only two atoms, and we saw how to count and place the electrons properly. Now, we will step up the complexity just a bit, and we will see molecules consisting of more than two atoms.

Let's start off by realizing that the level of complexity *cannot* get too high in these kinds of problems. To see why, let's go back to our Lego analogy. If I give you just a few Lego pieces, then there won't be so many ways to put these pieces together. In fact, I can choose pieces that will only fit together in one way. The same is true with molecules. If you have a molecule that is too big, there will be too many different structures that you can draw. For example, C_4H_3O will have too many different valid structures that can be drawn. And all of them are different compounds. So you won't ever be asked to draw the Lewis structure of C_4H_3O, because in the absence of more information, you will not be able to determine which structure to draw.

I mention this because it is comforting to realize that there is a limit to the complexity that you will see. The molecules you will draw will all be fairly small molecules. So, you just have to master a few simple rules for drawing Lewis structures.

In general, you will see molecules where there is one central atom. So, the first step is to select which atom in the molecular formula is the central atom. And that is what we will focus on in this section.

In some cases, it is very obvious, and you really don't need any tricks. For example, if you see $AlCl_3$, you would probably guess that Al is the central atom and the three chlorine atoms are attached to it. And you would be correct. But there are some cases where it is not so obvious. For example, COS (O is not the central atom here). Or, what about NO_2F? So, clearly, we need some tricks for choosing the central atom.

There are many tricks that you can use, but there is one general rule that will work in almost every case: always choose the *least electronegative atom*, but never choose hydrogen. We can easily rationalize why hydrogen can NEVER be the central atom. Hydrogen has one electron in its 1s orbital, and so, it can form only one bond. There are no other orbitals in the first energy level, so hydrogen does not have access to any other orbitals. So, it will never form more than one bond. Therefore, it will never be the central atom.

Let's see an example of how to choose the central atom. Consider the compound: $SOCl_2$. This compound has more than two elements (S, O, and Cl), so it is not so obvious to us which atom will be the central atom. So, we consult the periodic table, and we use the trends in electronegativity (that we discussed in Chapter 6) to determine which element is the least electronegative:

Increasing electronegativity →

H																	
Li	Be											B	C	N	**O**	F	
Na	Mg											Al	Si	P	**S**	**Cl**	
K	Ca	Sc	Ti	V	Cr	Mn	Fe	Co	Ni	Cu	Zn	Ga	Ge	As	Se	Br	
Rb	Sr	Y	Zr	Nb	Mo	Tc	Ru	Rh	Pd	Ag	Cd	In	Sn	Sb	Te	I	
Cs	Ba	La	Hf	Ta	W	Re	Os	Ir	Pt	Au	Hg	Tl	Pb	Bi	Po	At	

The trends indicate that sulfur will have the lowest electronegativity. Therefore, we choose sulfur to be our central atom. Let's see another example:

EXERCISE 7.13. Select the central atom in NO_2F.

Answer: There are three elements in this compound (N, O, and F). Find these elements on the periodic table to determine which one is the least electronegative:

Increasing electronegativity →

H																	
Li	Be											B	C	**N**	O	F	
Na	Mg											Al	Si	P	S	Cl	
K	Ca	Sc	Ti	V	Cr	Mn	Fe	Co	Ni	Cu	Zn	Ga	Ge	As	Se	Br	
Rb	Sr	Y	Zr	Nb	Mo	Tc	Ru	Rh	Pd	Ag	Cd	In	Sn	Sb	Te	I	
Cs	Ba	La	Hf	Ta	W	Re	Os	Ir	Pt	Au	Hg	Tl	Pb	Bi	Po	At	

You'll see that N is the least electronegative of these three elements, so choose N as your central atom.

As we said earlier, this rule (choosing the least electronegative element) will work in *almost* every case. But there is one common type of situation where this rule won't help. If we have to choose between C and S, we will be stuck. If we compare their positions on the periodic table, it is hard to determine (when we use the trends of electronegativity) which one is more electronegative:

Increasing electronegativity

H																	
Li	Be											B	C	N	O	F
Na	Mg											Al	Si	P	S	Cl
K	Ca	Sc	Ti	V	Cr	Mn	Fe	Co	Ni	Cu	Zn	Ga	Ge	As	Se	Br
Rb	Sr	Y	Zr	Nb	Mo	Tc	Ru	Rh	Pd	Ag	Cd	In	Sn	Sb	Te	I
Cs	Ba	La	Hf	Ta	W	Re	Os	Ir	Pt	Au	Hg	Tl	Pb	Bi	Po	At

So, in this case, we must resort to looking up the values in a table of electronegativity values. And when we do that, we realize that we are really stuck, because C and S have the same values. In this situation, you should choose C as the central atom (when you are comparing C and S). To understand this, you should appreciate that there are many different ways to measure electronegativity values. The values that you have in your textbook are the values that Linus Pauling measured with his method. But there are other methods that give different values (not shown in your textbook). If you were to look at these other methods for measuring electronegativity, you would find that C is less electronegative than S in most of these methods.

So, although Pauling's electronegativity values are not helpful to choose between C and S, you should generally treat C as being less electronegative than S.

Now let's get a little bit of practice. It's not really hard stuff. You can use the periodic table above, but you can also save yourself some time if you remember the two most electronegative elements: fluorine is the most electronegative, and oxygen is the second most electronegative element. If you remember that (and if you remember to never choose hydrogen as your central atom), then you will find that you will not have to access the periodic table to solve many of the following problems.

For each of the following compounds, select the central atom and place a circle around it:

7.14. SO_2F_2 **7.15.** $AlBr_2Cl$ **7.16.** CH_4S

7.17. COS **7.18.** H_2SO_4 **7.19.** $HClO_3$

7.20. $AlBr_3$ **7.21.** H_3SeO_4 **7.22.** IF_4

7.23. $HAsO_4$ **7.24.** BeI_2 **7.25.** H_3PO_3

7.4 DETERMINING THE SKELETON

Once we have selected the central atom, the next step is to determine how the rest of the atoms are connected to each other (the skeleton of the molecule). Once again, you will see examples where the answer is very obvious. When you only have two elements in the compound, and one of the elements has a subscript of 1 (for example, $AlCl_3$), then the skeleton seems obvious. You just connect the other atoms directly to the central atom:

$$
\begin{array}{c}
Cl \\
| \\
Cl{\diagdown}Al{\diagup}Cl
\end{array}
$$

In the example above, the skeleton was obvious, because there was a clear choice for a central atom, and the other elements nicely attached to the central atom. The choice was clear, because the compound only had two different kinds of elements (Al and Cl). And since one of the elements had a subscript of 1 (this is implied when there is no subscript shown), it was clear that we would just attach the chlorine atoms to the aluminum atom.

It is possible to have cases where neither element has a subscript of 1. In these cases, you probably won't have one central atom, but you can still figure out the skeleton if one of the elements is hydrogen. For example, consider N_2H_4. This compound has two elements and neither of them has a subscript of 1 in the molecular formula. Nevertheless, we can still figure out how to piece this together because we know that each hydrogen atom can only make one connection. So, the four hydrogen atoms must be at the ends of the molecule, and there is only one way for that to happen. The skeleton must be:

$$
\begin{array}{cc}
H & H \\
{\diagdown}N{-}N{\diagup} \\
H{\diagup} & {\diagdown}H
\end{array}
$$

So, in cases where the compound consists of only two kinds of elements (and one of those elements is hydrogen), it will be fairly obvious how to construct the skeleton. But what about compounds with *more* than two different kinds of elements? For example, consider the compound: $HClO_3$. We have seen how to select the central atom (it would be Cl in this case), but how do we choose the skeleton?

This compound actually belongs to a larger class of compounds that you should become familiar with. The molecular formula of these compounds will look like this: they will contain hydrogen, some other element, and then oxygen. Examples include HNO_2 or H_3PO_4. In both of these examples, you have hydrogen atoms in the front, some element in the middle, and then oxygen atoms at the end. These compounds are called oxoacids, and when you see a compound like this, you should place the oxygen atoms directly on the central atom, and then attach any hydrogen atoms to the oxygen atoms. Let's see an example of how this works:

EXERCISE 7.26. Draw the skeleton for H_2SO_4.

Answer: Always begin by selecting the central atom. Because you know that hydrogen will not be the central atom, the central atom must either be sulfur or oxygen. You can probably guess that it will be sulfur, because there is only one sulfur atom in the molecular formula, and there are four oxygen atoms. If you apply the rule (select the atom that is the least electronegative), you will see that the central atom will indeed be sulfur. You don't really need to look at a periodic table to know whether S or O is less electronegative. If you just remember that oxygen is the second most electronegative element (only second to fluorine), then you will know immediately that sulfur must be less electronegative.

Now you are ready to start drawing your skeleton. Begin by placing each of the oxygen atoms on the central atom:

$$
\begin{array}{c}
\text{O} \\
| \\
\text{O--S--O} \\
| \\
\text{O}
\end{array}
$$

Then, place the hydrogen atoms on two of the oxygen atoms:

$$
\begin{array}{c}
\text{O} \\
| \\
\text{H--O--S--O--H} \\
| \\
\text{O}
\end{array}
$$

(it doesn't matter which two oxygen atoms you pick). And, now you have your skeleton.

This is **not** a completed Lewis structure, because you still need to draw the rest of the electrons properly. For this particular example, you will soon see that you need to learn a few more concepts before you finish drawing the structure. So for now let's just focus on drawing the skeleton. You can concentrate on filling in the rest of the electrons in the upcoming sections.

There are rare examples where two oxygen atoms can be connected to each other. Two examples of such compounds are peroxides and per-acids:

$$
\text{R--}\boxed{\ddot{\text{O}}\text{--}\ddot{\text{O}}}\text{--R} \qquad \text{R--C--}\boxed{\ddot{\text{O}}\text{--}\ddot{\text{O}}}\text{--H}
$$

Peroxides   *Per-acids*

The letter R above refers to the rest of the molecule, which could be anything. The common feature of these compounds is the O—O bond. In Chapter 6, we talked about the oxidation state of oxygen atoms in cases like this, and we mentioned that these examples are fairly rare. The only example that is really common in general chemistry courses is hydrogen peroxide (H_2O_2):

$$
\text{H--}\ddot{\text{O}}\text{--}\ddot{\text{O}}\text{--H}
$$

Other than H_2O_2, when you see a compound with many oxygen atoms in the molecular formula, it is a pretty safe bet to assume that the oxygen atoms are **not** connected together. You should assume that each oxygen atom is probably connected to the central atom.

Let's use this concept in the following examples. Draw the skeleton for each of the following examples. Do not worry about completing the Lewis structure. For now, we are just focusing on drawing skeletons:

7.27. $HClO_3$ **7.28.** H_2SeO_3 **7.29.** $HBrO_4$

7.30. H_2CO_3 **7.31.** HNO_3 **7.32.** H_3PO_4

7.33. CO_3^{2-} **7.34.** NO_3^- **7.35.** $H_2PO_4^-$

7.36. HCO_3^- **7.37.** IO_3^- **7.38.** HSO_4^-

7.5   PLACING THE ELECTRONS

In the first section of this chapter, we practiced the general procedure for placing electrons in bonds and lone pairs. It went like this:

Start by counting the total number of valence electrons. Then count how many electrons are needed for every atom to have an octet. Based on the difference between these two numbers, you will know how many electrons need to be shared as bonds. Then place the rest as lone pairs, giving each atom an octet. In the end, don't forget to do a final check to make sure that you used the right number of electrons.

When you first practiced this procedure, you did it on simple compounds where you did not need to first determine the skeleton. Now, you will apply this method again to draw Lewis structures for compounds where you must first determine the skeleton. Let's see an example:

EXERCISE 7.39.   Draw the Lewis structure for COS.

Answer: Begin by selecting the central atom. Look for the least electronegative atom, and consult the periodic table to determine your answer:

															Increasing electronegativity →		
H																	
Li	Be											B	**C**	N	**O**	F	
Na	Mg											Al	Si	P	**S**	**Cl**	
K	Ca	Sc	Ti	V	Cr	Mn	Fe	Co	Ni	Cu	Zn	Ga	Ge	As	Se	Br	
Rb	Sr	Y	Zr	Nb	Mo	Tc	Ru	Rh	Pd	Ag	Cd	In	Sn	Sb	Te	I	
Cs	Ba	La	Hf	Ta	W	Re	Os	Ir	Pt	Au	Hg	Tl	Pb	Bi	Po	At	

From the periodic table, you can decide that C is the least electronegative atom, so choose C as your central atom (in Chapter 6 you learned that the electronegativity values of C and S are very close, but you should always choose C as the less electronegative element when you have a choice between C and S). Next, place the remaining atoms on the central atom, and you get:

$$S-C-O$$

Now you are ready to place all of the electrons in your structure. To do this, apply the method that you used in the beginning of this chapter. First count the number of valence electrons you have. Carbon has 4 valence electrons. Sulfur and oxygen each have 6 valence electrons. So, you have a total of $4 + 6 + 6 = 16$. If this were an ion, you would modify the number of electrons to account for the charge on the ion (like you saw earlier in this chapter). But this is a neutral molecule (no charge was indicated), so you have 16 valence electrons to work with.

Next, calculate how many electrons it would take to give every atom an octet (8 electrons for each atom). You have three atoms, so it would take $3 \times 8 = 24$ electrons.

To summarize, you need 24 electrons, and you only have 16. Therefore, you are missing 8 electrons. This tells you that you will need to share 8 electrons. So, you need a total of 4 bonds. The skeleton you chose already has 2 bonds in it, so you need 2 more bonds. So, place the two extra bonds, and you get:

$$S=C=O$$

So far, you have accounted for 8 electrons, but you have 8 more to place (because you started with 16 valence electrons). That means you need to place 4 lone pairs into your drawing. The carbon atom is already happy (it has 8 electrons around it). You can make the sulfur atom and oxygen atom happy, too, by giving each of them 2 lone pairs, and you get:

$$\ddot{\underset{..}{S}}=C=\ddot{\underset{..}{O}}$$

Don't forget to always double-check your answer. There count 16 electrons, and you started with 16 electrons, so you know that you did not accidentally forget some electrons (or put in too many).

Now, let's get some practice:

PROBLEMS.   For each of the following compounds, draw the Lewis structure.

7.40. H_2CO_3 **7.41.** HNO_2 **7.42.** $COCl_2$

7.43. NH_4^+ **7.44.** AlH_4^- **7.45.** PCl_4^+

7.46. CS_2 **7.47.** SiF_4 **7.48.** BF_4^-

7.6 ELEMENTS THAT VIOLATE THE OCTET RULE

In the beginning of this chapter, we discussed the octet rule, and we saw that certain elements will *violate* the octet rule. Recall the chart that we saw before:

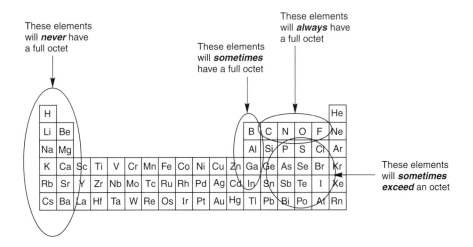

In the previous sections, we have seen how to draw Lewis structures for compounds containing elements that *always* follow the octet rule (C, N, O, F). Now, let's focus on the other three categories: (1) elements that *never* have an octet, (2) elements that *sometimes* have an octet, and (3) elements that sometimes *exceed* an octet.

Let's start with elements that *never* have an octet (the elements on the left side of the periodic table). These are the easiest to deal with, because you really won't have to draw Lewis structures of compounds containing these elements. A compound with one of these elements will likely be an ionic compound, and we usually don't use Lewis structures to show ionic bonds.

The next category includes elements that *sometimes* violate the octet rule. The elements in column 3A of the periodic table (see above) have three valence electrons, so we expect them to form compounds with three bonds. The most common examples are with boron and aluminum:

$$
\begin{array}{cc}
\overset{\displaystyle H}{\underset{H}{\overset{|}{\underset{\diagdown}{B}}}}{\diagup}^{H} &
\overset{\displaystyle H}{\underset{H}{\overset{|}{\underset{\diagdown}{Al}}}}{\diagup}^{H}
\end{array}
$$

We saw earlier in this chapter that these elements can form a fourth bond, but then they will have a formal charge of -1:

$$
\begin{array}{cc}
\overset{\displaystyle H}{\underset{\displaystyle H}{H{-}\overset{|}{\underset{|}{\overset{\ominus}{B}}}{-}H}} &
\overset{\displaystyle H}{\underset{\displaystyle H}{H{-}\overset{|}{\underset{|}{\overset{\ominus}{Al}}}{-}H}}
\end{array}
$$

In each of these cases, boron and aluminum have an octet, BUT now they are unhappy for another reason—they are negatively charged. So, we see that these elements (in column 3A) are always frustrated. In every situation, there is something making them unhappy (either not having an octet, or possessing a negative charge). This constant state of "frustration" defines much of the reactivity of these compounds. When you learn about acids and bases, you will see that these compounds (with three bonds, such as BCl_3 or $AlBr_3$) are prime examples of Lewis acids.

To draw compounds containing these elements (from column 3A), you will need to look at whether there are three or four bonds to the central element. You can determine this by looking to see if there is a charge of -1. If there is a charge, then you follow the exact same procedure that we learned before, because it obeys the octet rule. However, if there is no charge, you will need to make one minor modification. After you count how many valence electrons you have to work with, the next step is usually to count how many electrons it would take to give every atom an octet. That is where the change comes in. In AlF_3, you won't count the aluminum as needing 8 electrons. You will count it as if it only needs 6 electrons. Let's see this example more closely:

EXERCISE 7.49.   Draw the Lewis structure of AlF_3.

Answer: The central atom in this compound is aluminum, because it is less electronegative than fluorine. Aluminum belongs to column 3A of the periodic table. To determine if it will have an octet, look to see if it has a charge. Since no charge was indicated, you can conclude that aluminum only has three bonds in this compound, and it will **not** have an octet. You can verify from the molecular formula that aluminum only forms three bonds, because there are only three fluorine atoms. So, when you draw the skeleton you get:

$$
\overset{\displaystyle F}{\underset{F}{\overset{|}{\underset{\diagdown}{Al}}}}{\diagup}^{F}
$$

Next, count the number of valence electrons that you have to work with. Aluminum has three valence electrons and each fluorine atom has seven valence electrons. So, you have a total of $3 + 3(7) = 24$ electrons.

Next, count how many electrons you will need to give every atom an octet. BUT WAIT. This is where you modify the procedure. You know that the fluorine atoms will each get 8 electrons, *but* you have already concluded that the aluminum atom will *not* have an octet in this case. It will only have 6 electrons around it, rather than 8. So, you *don't* need 4×8 electrons. Rather, you only need 6 electrons for the aluminum atom and 8 electrons for each fluorine atom. So, you only need a total of $6 + 3(8) = 30$ electrons to accurately draw the Lewis structure of this compound. This is actually the way you treated cases involving hydrogen—you treated the hydrogen as if it only needed 2 electrons instead of 8. Similarly, here the aluminum atom is treated as if it only needs 6 electrons instead of 8.

To summarize, you have seen that you need 30 electrons, and you only have 24. That means that you will need to share 6 electrons. So you need to have 3 bonds. Your skeleton already has 3 bonds in it, so now you just need to place the remaining electrons as lone pairs. You started with 24 electrons, and you have already used 6 of them. So, you have 18 more to place. 18 electrons = 9 lone pairs. Place electrons in a way that gives every fluorine atom an octet, and you know that aluminum is NOT going to have an octet in the end:

And that is your final answer.

PROBLEMS.   Draw the Lewis structure for each of the following compounds.

7.50. BCl_3 **7.51.** $AlCl_4{}^-$ **7.52.** BH_3

Now, we just have one more category to focus on. We need to talk about elements that sometimes *exceed* an octet:

H																	He
Li	Be											B	C	N	O	F	Ne
Na	Mg											Al	Si	P	S	Cl	Ar
K	Ca	Sc	Ti	V	Cr	Mn	Fe	Co	Ni	Cu	Zn	Ga	Ge	As	Se	Br	Kr
Rb	Sr	Y	Zr	Nb	Mo	Tc	Ru	Rh	Pd	Ag	Cd	In	Sn	Sb	Te	I	Xe
Cs	Ba	La	Hf	Ta	W	Re	Os	Ir	Pt	Au	Hg	Tl	Pb	Bi	Po	At	Rn

These elements will **sometimes** **exceed** an octet

The *most common* examples of elements that exhibit expanded octets are S, P, Cl, Br, and I. Let's pick one, and take a close look at it. Consider the compound: PCl_5. When we draw the Lewis structure for this compound, we start by drawing the skeleton, and we get:

$$\begin{array}{c} Cl \\ | \\ Cl-P-Cl \\ / \backslash \\ Cl \quad Cl \end{array}$$

Immediately, we notice a problem. Phosphorous has five bonds (or ten electrons). That means that it is exceeding its octet. We have seen that second-row elements will NEVER do this. Second-row elements will NEVER form five bonds. But third-row elements have other orbitals that can be accessed (this topic is taken up in greater depth in Chapter 8, when we discuss hybridization). So, it is not a problem for phosphorous to have five bonds. In this example, phosphorous exceeds its octet.

There are other examples where it is not so clear whether or not we should draw a structure that exceeds the octet rule. For example, consider the following compound: $HClO_3$. We have actually seen this compound before. When we were learning how to choose skeletons, we saw this specific example (see Problem 7.27). We saw that the skeleton was:

$$\begin{array}{c} H \\ | \\ O \\ | \\ O-Cl-O \end{array}$$

In order to fill in the rest of the electrons, we will apply the method that we used in the beginning of this chapter. We start by counting the total number of valence electrons. Then we count how many electrons are needed for every atom to have an octet. Based on the difference between these two numbers, that tells us how many electrons need to be shared as bonds. Then we place the rest as lone pairs in such a way that gives each atom an octet, and we end up with:

H
|
:O:
|
:O–Cl–O:

Now, let's analyze each atom to see if it has a formal charge. When we do, we see the following formal charges (if you forgot how to assign formal charges, you should definitely go back to Chapter 6 and review the procedure, because you will need that skill now):

H
|
:O:
|
:O–Cl–O:

There are a lot of formal charges on this drawing, and there is a way to get rid of them. When you have a positive charge and a negative charge next to each other, you can usually eliminate both charges by moving a lone pair to form an extra bond:

H   H
| |
:O: :O:
| ⟶ |
:O–Cl–O: O=Cl=O

This gets rid of our formal charges, but now we see that chlorine is violating the octet rule. We know that chlorine is *allowed* to violate the octet rule, but how do we know which structure to draw? Do we draw the one with formal charges that does not violate the octet rule? Or do we draw the one with no formal charges that violates the octet rule? Unfortunately, this is somewhat of a debate. Theoretical chemists argue about this very issue. And that debate has filtered down into the textbooks. Some textbooks will instruct you to minimize formal charges and violate the octet rule. Other textbooks will instruct you to keep the formal charges and not to violate the octet rule. You should look in your textbook to see which side your textbook takes (and you should pay attention in class to see which side your instructor takes).

If you belong to the camp that says you should violate the octet rule in order to minimize formal charges, then BE CAREFUL. You can NEVER do this with a second-row element. For example, consider the structure of nitric acid, HNO_3. If you go through the steps to draw the Lewis structure for nitric acid, you will find that the structure is:

:O:
||
:O–N–O–H

Notice that the nitrogen atom obeys the octet rule, but there are formal charges. We might be tempted to pull down a lone pair to get rid of the formal charges (if we belong to the camp that says to do so):

$$:\!\overset{\ominus}{\underset{}{\text{O}}}\!\!-\!\!\overset{:\text{O}:}{\underset{\oplus}{\text{N}}}\!\!-\!\!\overset{}{\underset{}{\text{O}}}\!\!-\!\!\text{H} \quad\longrightarrow\quad :\!\text{O}\!\!=\!\!\overset{:\text{O}:}{\underset{}{\text{N}}}\!\!-\!\!\overset{}{\underset{}{\text{O}}}\!\!-\!\!\text{H}$$

But we don't do this, because that would give the nitrogen atom five bonds. And nitrogen is a second-row element. We have seen that second-row elements NEVER have five bonds. They simply cannot EVER do this. So, we leave it the way it was, and we do not violate the octet rule (according to both camps):

$$:\!\overset{\ominus}{\underset{}{\text{O}}}\!\!-\!\!\overset{:\text{O}:}{\underset{\oplus}{\text{N}}}\!\!-\!\!\overset{}{\underset{}{\text{O}}}\!\!-\!\!\text{H}$$

In fact, this example clearly points out the potential pitfall with saying that it is OK to *sometimes violate* the octet rule. That is a dangerous statement to make, because it can confuse you. Nitrogen will NEVER form five bonds. That kind of violation will NEVER happen.

To recap, we have seen two different methods (violate the octet rule in order to get rid of the formal charges, or keep the formal charges in order to avoid violating the octet rule). Let's see a worked example first where we use both methods. We will see that the two methods are the same up until the very end of the problem:

EXERCISE 7.53. Draw the Lewis structure of $HBrO_2$.

Answer: Begin by selecting the central atom. You know that you should never choose hydrogen as the central atom. Between Br and O, you have to choose Br as the central atom. Both bromine and oxygen are very electronegative elements, but you know that oxygen is more electronegative (because oxygen is the second most electronegative element, after fluorine). Remember to place the hydrogen atom on one of the oxygen atoms, and you get the following skeleton:

$$\text{O}-\text{Br}-\text{O}-\text{H}$$

Now you are ready to place all of the electrons in your structure. First count the number of valence electrons you have. Bromine has 7 valence electrons, each oxygen has 6 valence electrons, and hydrogen has 1. That gives you $(7) + 2(6) + 1 = 20$.

Next, calculate how many electrons it would take to give every atom an octet. The bromine atom needs 8, each of the oxygen atoms need 8, and the hydrogen needs 2 electrons. So you need $(8) + 2(8) + (2) = 26$ electrons.

To summarize, you need 26, and you only have 20. Therefore, you are missing 6 electrons. This tells you that you will need to share 6 electrons. So, you need a total of 3 bonds. The skeleton you chose already has 3 bonds in it, so now you need to place the rest of the electrons as lone pairs. So far, you have accounted for 6 electrons, but you have 14 more to place (because you started with 20 valence electrons). That means you need to place 7 lone pairs, and you do this in such a way that gives each element an octet (with hydrogen getting 2 electrons):

$$:\ddot{O}-\ddot{Br}-\ddot{O}-H$$

Next, calculate the formal charges, and you get:

$$^{\ominus}:\ddot{O}-\overset{\oplus}{\ddot{Br}}-\ddot{O}-H$$

There are two formal charges; one positive charge and one negative charge.

This is the point where the two methods become different: One method will tell you to leave it the way it is (keeping the formal charges and not violating the octet rule). But the other method will tell you to get rid of the formal charges by drawing a double bond:

$$^{\ominus}:\ddot{O}-\overset{\oplus}{\ddot{Br}}-\ddot{O}-H \longrightarrow \ddot{O}=\ddot{Br}-\ddot{O}-H$$

So, you can see that there are two possible answers to this question, depending on which argument you stick with. The two possible answers are:

$$^{\ominus}:\ddot{O}-\overset{\oplus}{\ddot{Br}}-\ddot{O}-H \qquad \ddot{O}=\ddot{Br}-\ddot{O}-H$$

Your textbook will advocate one of these drawings as the more correct answer. You should make sure to find out which method you are expected to use.

PROBLEMS. For each of the problems below, there will be two possible answers (depending on which method you use). You should first look in your textbook and lecture notes, and determine which method you have been instructed to use. Then, draw the Lewis structure for each of the following compounds using that method.

7.54. $HClO_2$ **7.55.** HIO_3 **7.56.** SO_2

7.57. SO_3 **7.58.** $HBrO_4$ **7.59.** $POCl_3$

      We have seen two methods that seem to be at odds with each other (violate the octet rule in order to get rid of the formal charges, or keep the formal charges in order to avoid violating the octet rule). In the next section, we will explore a topic called *resonance*. When we are done with that section, we will be in a much better position to understand the nature of the argument that we have just seen (the two different methods).

7.7   RESONANCE STRUCTURES

In this chapter, we have seen how to draw Lewis structures. However, these drawings have one MAJOR deficiency. Although our drawings are very good at showing which atoms are connected to each other (the skeleton), our drawings are ***not good*** at showing where all of the electrons are. We have seen this explicitly in the debate that we discussed in the end of the previous section. To understand the issue at hand, we need to think about how we defined electrons earlier.

      Electrons aren't really solid particles that can be in one place at one time. All of our drawing methods treat electrons as particles that can be placed in specific locations (in Lewis structures, we draw them as dots located in specific locations). But this is not accurate. It is best to think of electrons as ***clouds of electron density***. We saw this analogy in Chapter 6. We said that electrons are not flying around *in* clouds; but rather, they *are* clouds. These clouds can sometimes spread themselves across large regions of a molecule. And that is the problem. When an electron cloud is spread out over a large region of the molecule, how are we supposed to draw that? Lewis structures assume that electrons are little dots whose location can be either here or there. But that is not a correct assumption.

So how do we represent a molecule if we can't draw where electrons are? The answer is *resonance*. We use the term *resonance* to describe our solution to the problem: we use *more than one* drawing to represent a single molecule. We draw several drawings, and we call these drawings *resonance structures*. We meld these drawings into one image *in our minds*. To better understand how this works, consider the following analogy.

Your friend asks you to describe what a nectarine looks like, because he has never seen one before. You aren't a very good artist, so you say the following:

> Picture a **peach** in your mind, and now picture a **plum** in your mind.
> Well, a **nectarine** has features of both: the inside tastes like a peach,
> but the outside is smooth like a plum. So, take your image of a peach
> together with your image of a plum and **meld them together** in your
> mind into one image. That's a nectarine.

It is important to realize that a nectarine *does not* switch back and forth every second from being a peach to being a plum. A nectarine is a nectarine, *all of the time*. The image of a peach is *not adequate* to describe a nectarine. Neither is the image of a plum. But by imagining both together at the same time, you can get a sense of what a nectarine looks like.

The problem with drawing molecules is similar to the problem with the nectarine. For some molecules, no one drawing adequately describes the nature of the electron density spread out over the molecule. To deal with this problem, we make several drawings, and then we **meld them together** in our mind into one image—just like the nectarine.

Let's see an example:

The ion above has two important resonance structures. Notice that we separate resonance structures with a two-headed arrow, and we place brackets around the structures. The arrow and brackets indicate that these are two ways to draw *one molecule*. The molecule is *not* flipping back and forth between the different resonance structures. The electron density in the molecule is *not* shifting around.

Resonance often confuses students. It is tempting to think that something is actually happening here. The term "resonance" means something to all of us before we started learning chemistry (tuning forks causing each other to vibrate because of *resonance*), so we naturally want to think that resonance in chemistry means that the molecules are "resonating". But they are not. They are not doing anything at all.

Resonance is just the term that chemists use to describe the fact that Lewis structures are inadequate drawings, because these drawings assume that electrons can be in very well-defined locations. But we have seen that electrons should really be thought of as clouds that can sometimes be spread out over a large region of a molecule. That would be too difficult to draw, and most chemists are not artists. So, to deal with the inadequacy of our drawings, we make several drawings and we meld them together in our minds. Then we call this concept "resonance". It is definitely a bad (and confusing) term.

To make matters worse, we talk about a term called "resonance energy". This might cause you to think that the molecule is doing something that puts it in a lower energy state. But, once again, the molecule isn't doing anything. The term *resonance energy* is used to describe the stability associated with having electrons that are spread out over a large region of the molecule. Remember that electrons have a negative charge, so spreading the charge is a good thing. But when we draw individual Lewis structures, each Lewis structure will not individually capture this information.

Let's go back to the example we just saw:

The ion shown above cannot be drawn with just one structure, because the negative charge is **not** located on just one oxygen atom. Rather, there is an electron cloud that is *spread out* over both oxygen atoms. There is no single drawing that will capture this information, so we draw two drawings in order to get a better picture of where the charge is spread out. When we draw these Lewis structures and meld them together in our minds, we need to also think about the stability associated with a spread-out electron cloud. This stability is referred to as *resonance energy*. Once again, the terminology can be misleading. It is so easy to misinterpret this expression and believe that the molecule is *doing* something that is lowering its energy. It is not. The term *resonance energy* is just describing the fact that one of the electron clouds is spread out.

Another confusing issue is the double-headed arrow that we use to show resonance structures:

This has been specifically chosen by chemists to look different than an equilibrium arrow which shows a reaction taking place:

Chemists have already trained their eyes to recognize the difference between these two types of arrows, but students often miss the difference. These arrows mean totally different things. We use the first arrow to show that we are drawing resonance structures. We use the second arrow to show that a reaction is actually taking place. The similarity of these arrows is another reason why students often get confused and incorrectly believe that resonance is something that is happening.

Now that we have a better understanding of what resonance is, we must learn how to draw resonance structures. There are many rules that are used for drawing resonance structures. The topic of resonance will become much more important when you begin studying organic chemistry next year. You will see that there are many different patterns that you need to learn in order to draw resonance structures (you can see instructions for each of these patterns in *Organic Chemistry as a Second Language*). Fortunately, only one of these patterns is commonly seen in a general chemistry course, so we will focus on this pattern now.

There are many examples that all exhibit this same pattern. These compounds will generally be negatively charged ions that have multiple oxygen atoms (NO_3^-, SO_4^{2-}, CO_3^{2-}, PO_4^{3-}, etc.). Let's focus on a specific example. Consider the structure of NO_3^-. If we apply the method for drawing this compound, we will find that there are three equivalent drawings that we can make:

Remember that we never draw five bonds to nitrogen. So, we can agree that we should not get rid of the formal charges in the structures above (as described in the previous section). These drawings all represent one molecule that has an electron cloud spread out over all three oxygen atoms.

If you had tried to draw NO_3^- (using the process we learned in this chapter), then you would have arrived at a point where you wouldn't know where to put all of the electrons. And that is OK. That is what happens when you try to draw the Lewis structure of a compound that has resonance structures. To see how this works, let's actually do this example step-by-step:

EXERCISE 7.60. Draw all resonance structures for the nitrate ion (NO_3^-).

Answer: Begin by selecting the central atom. Nitrogen is less electronegative than oxygen, so choose nitrogen as your central atom. That gives you the following skeleton:

$$O-N-O \text{ with } O \text{ above}$$

Now you are ready to place all of the electrons in your structure. First count the number of valence electrons you have. Nitrogen has 5 valence electrons and oxygen has 6 valence electrons. That gives you $(5) + 3(6) = 23$. Don't forget the negative charge which means that you have one extra electron. So, in total, you have **24** electrons to work with.

Next, calculate how many electrons it would take to given every atom an octet. Since you have four atoms, it would take $4 \times 8 = \boldsymbol{32}$ electrons.

To summarize, you need 32 electrons, and you only have 24. Therefore, we are missing 8 electrons. This tells you that you will need to share 8 electrons. So, you need a total of 4 bonds. The skeleton you chose already has 3 bonds in it, so you need to place one more bond. And now you are stuck. Where should it go? Is there any rational reason to choose one location over another? No, there isn't. This should be a sign to you that there will be resonance structures here. There are three different places where you could put the bond, so draw all three possibilities:

So far, you have accounted for 8 electrons, but you have 16 more to place (because we started with 24 valence electrons). That means you need to place 8 lone pairs into each of your drawings. You can do this in such a way that it gives every atom an octet, and you get:

Finally, draw in the resonance arrows, place brackets around the whole thing, and draw in the formal charges:

And now you have drawn all of the resonance structures. These structures have formal charges that you cannot get rid of, because nitrogen cannot have more than four bonds. Remember that nitrogen is a second-row element, so it can NEVER have five bonds.

Now let's get some practice. This will be the final set of problems for this chapter. In each of the following problems, draw the resonance structures for the ion shown.

7.61. NO_2^-

7.62. CO_3^{2-}

Now we can more fully understand the nature of the debate that was described in the previous section. Recall that you looked at the Lewis structure for $HClO_3$, and you saw that there were two possible answers:

The difference between these answers comes from which method you chose to use. The first drawing came from using the method where you keep the formal charges (in order to avoid violating the octet rule). The second drawing came from using the method where you get rid of the formal charges (by violating the octet rule).

Now that you know what resonance is, you are in a better position to understand the nature of this debate. The two drawings above are really just resonance structures.

You might ask: if that is the case, then why are theoretical chemists arguing with each other over which structure is more correct? Can't we just say that both drawings are correct?

Now we can understand this argument in a new light. Let's go back to the nectarine analogy (earlier in this section). A nectarine is not a peach, and it is not a plum. But we got it from mixing the two together. Now imagine I ask you the following question: does a nectarine look more like a peach or more like a plum? You would probably answer that it looks more like a peach. In other words, although we mixed a peach and plum to get a nectarine, it is clear that the nectarine's "peach character" is stronger than its "plum character".

We can apply the same question to the drawings above. The two drawings above are actually resonance structures. That means that if we really want to understand the nature of the molecule, we will need to draw both drawings, and then meld them together in our minds. But let's say we ask the following question: does the molecule look more like the first drawing or more like the second drawing? And that is where the debate comes in. Chemists don't argue over whether the two drawings are resonance structures. They argue about the "character" of the molecule. Does it have more character of the first drawing, or more character of the second drawing?

You would probably argue that a nectarine looks more like a peach than a plum (and I would agree with you). But if someone were to tell you that a nectarine actually looks more like a plum, you would probably end up embroiled in an argument. And that is the type of argument that we saw in the previous section.

CHAPTER 8

MOLECULAR GEOMETRY AND HYBRIDIZATION STATES

In this chapter, we will focus on the individual steps involved in predicting the shapes of small molecules. These skills will be very important, not only for helping you solve exam problems that ask you to predict molecular shape, but also for helping you understand other principles in general chemistry. For example, shape is the basis for polarity, which is very important for the properties of phases and mixtures. Shape is also important for reactivity and mechanisms. Throughout your textbook, you will certainly see at least a few examples where the shape of a molecule *changes* during a reaction. When you analyze how the shape of a molecule changes during a reaction, you can better understand the factors that influence the rate of the reaction. For example, is the transition state very high energy or not? These are all important considerations, and they are all dependent on a solid understanding of geometry. So, it is certainly worth your time to learn how to predict the geometry of molecules.

It becomes a lot easier once you realize that the shape of most small molecules can be predicted by one simple concept: electrons repel each other. Let's take a closer look.

8.1 ORBITALS AND THEIR GEOMETRY

I recently purchased a toy construction set (called GeoMags) for my children. The set has only two kinds of pieces: magnetic bars and metal spheres:

By attaching many magnets and spheres, you can construct all kinds of interesting shapes. I must admit that I found myself playing with these toys more than my children were. I found myself mesmerized by the ability of magnets to repel each other. It is a very strange sensation when you try to force the negative ends of two magnets together. You can *feel them* repelling each other, even though there is apparently nothing in between the two magnets. It is almost magical. As a side note, it is not entirely accurate to refer to the magnets as having "negative ends". This is an oversimplification, but I believe it can help us learn how to predict the shapes of molecules.

When I was playing with these magnets, I noticed something interesting (and somewhat predictable). I began to attach several magnets to one metal sphere, and I made sure that the *negative end* (so to speak) of each magnet was pointing away from the sphere:

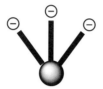

When I placed three magnets on the sphere, I noticed that the magnets repelled each other, and they began to slide along the surface of the sphere in an attempt to get farther apart from each other:

I was curious to see what would happen as I continued to attach more magnets to the sphere. When I placed **six** magnets on the sphere, they seemed to move toward forming a special arrangement called an octahedron:

In an octahedral arrangement, the six magnets achieve a maximum distance from one another (with 90° angles separating any two neighboring magnets). I then tried to push the magnets closer together with my fingers, by sliding each magnet along the surface of the sphere:

But as soon as I let go, the magnets would slide away from each other again. This should make a lot of sense. The magnets were just repelling each other.

When I used four magnets instead of six, I obviously did not get an octahedron. I got a different arrangement, but this new arrangement was still determined by the same principle (the magnets were repelling each other).

There are *five* important molecular arrangements that we need to be familiar with, and we can use these magnetic toys to "see" the five arrangements. We can generate these five different arrangements by attaching anywhere from two to six magnets to a sphere, and applying the all-important principle that the magnets repel each other.

Let's start with the first arrangement, which we will get when we place two magnets on a sphere:

This arrangement is quite simple. It should be obvious that the magnets achieve a maximum distance from each other when they are separated by 180°. We call this a *linear* arrangement, because the two bonds form a straight line.

Next, let's consider what arrangement we get when we put three magnets on a sphere. In order for all three magnets to get as far apart as possible, they arrange themselves like this:

Notice that all three magnets are in the same plane, and the bond angles are 120°. Therefore, we call this arrangement *trigonal planar*.

When we attach four magnets to a sphere, they achieve maximum distance from each other like this:

You can think of this structure like a tripod (like the ones used for cameras): three magnets are being used as legs, and the fourth magnet is pointing straight up. In this arrangement, all of the bond angles are identical (approximately 109.5°). This is called a *tetrahedral* arrangement.

When we put five magnets on a sphere, we get a new arrangement that is somewhat hard to remember. So, you might want to stare at this arrangement for a while.

Three of the magnets are arranged in a plane (the three that are resting on my hand), and the remaining two magnets are pointing straight up and straight down. Here is another view of the same arrangement:

When we try to capture this arrangement in a drawing, we draw it like this:

This arrangement is called **_trigonal bipyramidal_**. To understand why, imagine drawing lines connecting all of the spheres (except for the central sphere):

These lines sketch out a structure that looks like two pyramids stuck together. If you have trouble seeing the two pyramids, then think of it like this. Imagine a triangle that serves as the base for two pyramids. The top pyramid is shown with thin black lines, the bottom pyramid is shown in gray lines, and the common base for these two pyramids is shown in bold lines:

"Bipyramidal" indicates that it looks like two pyramids stuck together, and "trigonal" indicates that the common base has three sides.

Finally, the last arrangement we need to consider is when we put **_six_** magnets on a sphere. We have already seen this arrangement before, and we called it **_octahedral_**.

To understand why we call this an **_octahedral_** shape, we will use the same method that we used just a moment ago. We draw lines connecting all of the spheres (except for the central sphere), and we see a shape that, once again, looks like two pyramids stuck together. But this time, the base of the pyramids has _four_ sides (rather than three sides, like in the trigonal bipyramidal shape above):

Perhaps it might make sense to call this shape *square bipyramidal*, but that's not what we call it. Instead, we look at the object, and we see that it has a total of eight faces (four faces on the top pyramid and four faces on the bottom pyramid). Since it has **eight** faces, we call it **oct**ahedral.

So, all in all, we have seen five basic arrangements: linear, trigonal planar, tetrahedral, trigonal bipyramidal, and octahedral. In every case the arrangement was determined by the number of magnets that we attached to the sphere. All we need to know is the number of magnets that we used, and we can predict the geometry (because we know that the magnets will repel each other, and will want to get as far apart as possible). Let's summarize the five basic arrangements that we have seen:

# of Magnets	Arrangement	Illustration
2	Linear	
3	Trigonal planar	
4	Tetrahedral	
5	Trigonal bipyramidal	
6	Octahedral	

When it comes to molecules, we will use the exact same logic. All we need to know is the number of occupied *orbitals* in the valence shell. These orbitals will always be as far apart as possible (because orbitals are filled with electron density; and electrons repel each other). This concept is called "Valence Shell Electron Pair

Repulsion Theory" (or VSEPR for short). The basic idea is that all bonds will want to be as far apart as possible—and that determines the geometry.

We will soon see that some orbitals can be filled with lone pairs (rather than being used to form bonds), and we will have to see how this affects the geometry around an atom. For now, let's just focus on small molecules where the central atom has *no lone pairs*. Once we are comfortable with the five basic arrangements, then we can see how these arrangements get modified by the presence of lone pairs.

EXERCISE 8.1. Predict the shape of PCl_5.

Answer: Clearly, P is the central atom, surrounded by five chlorine atoms. So, you'll need to focus on the geometry of the phosphorus atom. That will determine the shape of this molecule.

P is in column 5A of the periodic table. Therefore, it has *five* valence electrons that can be used to form bonds or lone pairs. In this example, there are five bonds, so P is using all five of its valence electrons to form bonds. That means there are no lone pairs. The five orbitals being used to form these five bonds will all want to be as far apart as possible. We have seen that five groups can achieve a maximum distance from each other when they are arranged in a *trigonal bipyramidal* shape.

Before doing some problems yourself, go back and review the chart on page 238. Make sure that you know the names of all five arrangements.

For each of the following compounds, predict the geometry. Remember that it will be one of the five basic arrangements that we discussed. Just count how many groups need to be as far apart as possible. Try to do these without looking back at the previous pages. Try to remember what the names are—that is the whole point of these problems:

8.2. $BeCl_2$ **8.3.** $AlCl_3$ **8.4.** CCl_4

8.5. SF_6 **8.6.** PBr_5 **8.7.** BF_3

8.2 ELECTRON DOMAINS

In all of the examples above, the orbitals were only used to form bonds. There were no examples of a central atom that had lone pairs. Now we need to consider what effect lone pairs will have on predicting shape. In order to do this, let's quickly review orbitals:

We saw in Chapter 6 that orbitals can either (a) be empty, (b) contain one electron, or (c) contain two electrons. These are the only possibilities for any orbital. When an orbital is empty, then it is just empty space, and there is nothing to talk about. So we just have to consider the other two possibilities: an orbital with one electron, and an orbital with two electrons.

When an orbital has one electron, it will generally be used to form a bond by overlapping with an orbital of another atom (see Chapter 6 for a better description of what a bond is). It is possible for an orbital to have one electron that is **not** used to form a bond. Compounds containing orbitals like this (with only one electron) are called radicals, and they are usually very unstable (they react very rapidly). There are a few examples of stable, ground-state diradicals (such as molecular oxygen—O_2). You will see these examples when you learn about paramagnetism. For now, we will not worry about these examples, because you will rarely have to predict the geometry of a compound where this needs to be taken into account.

When an orbital has two electrons, it is called a lone pair. We saw how to draw these lone pairs when we drew Lewis structures (see Chapter 7).

So, to summarize what we have seen about orbitals, we really only have two *common* uses for orbitals. Almost every orbital will be used either to form *a bond* or to contain *a lone pair*. In the examples we saw in the previous section, we focused on molecules where the central atom had no lone pairs. Now we need to expand our horizons a bit. But we won't make the issue any more complicated. The basic idea is still the same: all orbitals (whether they contain lone pairs, or whether they are used to form bonds) will repel each other, and will therefore arrange themselves so that they are as far apart as possible.

So, we see that we treat lone pairs just like bonds when we are trying to predict how the orbitals will arrange themselves in space. Therefore, from now on, we will use new terminology that many textbooks have begun to use—we will call the orbitals *electron domains*. An *electron domain* is really just an orbital that is being used (either to form a bond or a lone pair). Therefore, a bond is called an electron domain, *and* a lone pair is also called an electron domain. So, the idea goes like this: always count up the total number of electron domains and that will tell you how these domains (orbitals) will arrange themselves. For example, if we count that an atom has three electron domains, then the arrangement of the electron domains will be trigonal planar. If we count that an atom has five electron domains, then the arrangement of the electron domains will be trigonal bipyramidal.

Notice that we have not yet used the word geometry, or molecular shape. It actually turns out that the geometry of a molecule will often be different than the arrangement of the electron domains (and we will see exactly how in just a few moments). But in order to predict the geometry for any molecule, you must always first determine the arrangement of electron domains around the central atom. That is the first step. Let's see an example of how to do this:

EXERCISE 8.8.   Determine the arrangement of the electron domains in the central atom of ammonia (NH_3).

Answer: Begin by drawing the Lewis structure, because you will need to know how many bonds and how many lone pairs are on the central atom. When using the method for drawing Lewis structures (see Chapter 7 if you are rusty on Lewis structures), you get the following:

$$H-\overset{\displaystyle ..}{\underset{\displaystyle |}{N}}-H$$
$$H$$

The nitrogen atom is the central atom, and it has *three bonds* and *one lone pair*. Therefore, it has a total of *four* electron domains. And you have seen that four electron domains can get the farthest apart when they are in a *tetrahedral* arrangement.

The method being used here is exactly the same as the method used in the previous section (where the examples had no lone pairs). We are just broadening the method now by treating lone pairs as electron domains as well. Now, we are counting bonds *and* lone pairs.

We are almost ready to learn how to predict *molecular shapes*. But first, let's make sure that we are comfortable determining the *arrangement of electron domains*. For each of the following compounds, determine the arrangement of electron domains around the central atom. You will have to start by drawing the Lewis structure so that you can see how many bonds and lone pairs are on the central atom.

Then, use that total number (bonds + lone pairs) to determine the arrangement of electron domains around the central atom:

8.9.　SBr_2

Lewis structure = _____

What do you predict will be the arrangement of the electron domains in the compound above:

| Linear | Trigonal planar | Tetrahedral | Trigonal bipyramidal | Octahedral |

8.10.　ICl_3

Lewis structure = _____

What do you predict will be the arrangement of the electron domains in the compound above:

| Linear | Trigonal planar | Tetrahedral | Trigonal bipyramidal | Octahedral |

Let's do just a few more problems. Remember how we solve these problems: first draw out the Lewis structure, and then count the bonds *and* lone pairs on the central atom. Your answer should be one of the five basic arrangements that we have learned (try to do these problems without looking above at the previous problems to see the five possible answers—you need to get to a point where you know the five possible arrangements without someone showing them as a multiple choice problem).

8.11. What do you *predict* will be the arrangement of the electron domains in PBr_3?

8.12. What do you *predict* will be the arrangement of the electron domains in BrF_5?

8.13. What do you *predict* will be the arrangement of the electron domains in NH_4^+?
(If the positive charge throws you off, then you might want to review the section in Chapter 7 where we discussed how to draw the Lewis structures of ions.)

8.3   PREDICTING MOLECULAR GEOMETRY

In the previous section, we learned how to count the total number of electron domains (lone pairs + bonds), and then we used that number to determine the arrangement of those electron domains.

Now we are ready to use that information to predict the shapes of small molecules. When we talk about the shape of a molecule (or its geometry), we are *not* referring to the arrangement of the electron domains; rather, we are referring to the geometry of the *atoms* that are connected to each other. Luckily, this geometry is determined by the arrangement of the electron domains. As an example, let's revisit the structure of ammonia (NH_3) again.

In the previous section, we saw that the central nitrogen atom in ammonia has four electron domains (three bonds and one lone pair). Therefore, the arrangement of the electron domains is tetrahedral:

But take a close look at the tetrahedral arrangement shown above. Notice that one of the electron domains appears to be different than the others (the one pointing straight up). This electron domain represents the lone pair. So, in the picture above, we can clearly see all three bonds *and* the lone pair. And we see that these four electron domains are in a tetrahedral arrangement. BUT, when we talk about the *shape* of a molecule, we are referring to the connectivity of *atoms*, and we ignore the lone pairs for a moment. If we only focus on the atoms, we will see the shape of a pyramid, with a three-sided base:

So now we have a new shape, and we call it *trigonal pyramidal*, even though the arrangement of the orbitals is still tetrahedral. You should appreciate that this new shape is *based on* one of the original five arrangements (the tetrahedral arrangement). In fact, that is the effect that lone pairs have. Once we have determined the arrangement of the electron domains (which will always be one of the five basic arrangements that we learned), then we take the lone pairs into account. And that gives us the final molecular shape.

Before looking at another example, let's review the overall strategy when determining the shape of a small molecule. We begin by determining which atom is the central atom, and then we draw the Lewis structure. The Lewis structure shows us how many electron domains the central atom has (we count the bonds and the lone pairs—the total number will either be 2, 3, 4, 5, or 6). Based on that number, we then determine the arrangement of the electron domains (which will be one of the five basic arrangements that we discussed: 2 = linear, 3 = trigonal planar, 4 = tetrahedral, 5 = trigonal bipyramidal, and 6 = octahedral). Finally, we determine how many of the electron domains are lone pairs, and then we take this into account by ignoring the lone pairs for a moment and focusing only on the bonds. Let's see another example:

EXERCISE 8.14.   Determine the geometry of H_2O.

Answer: You will need to draw the Lewis structure, because you need to know how many bonds and lone pairs are on the central atom. Follow the method we described in Chapter 7, and you'll get the following Lewis structure:

$$H-\overset{..}{\underset{..}{O}}-H$$

Notice that the central atom has two bonds and two lone pairs.

Be careful here. Don't let the drawing above fool you into thinking the molecule is linear. It is NOT. Lewis structures do NOT show geometry. They just show how many bonds and how many lone pairs there are. Lewis structures make no attempt to depict the three-dimensionality of the molecule.

If we want to determine the shape, we will need to use the Lewis structure to count the number of electron domains on the central atom. In this example, the central atom has two bonds and two lone pairs, so it is using a total of *four* electron domains. These four electron domains will all want to be as far apart as possible, so they will be arranged in a *tetrahedral* arrangement:

Finally, you need to take the lone pairs into account. Although the electron domain arrangement is tetrahedral, the geometry of the molecule only refers to the bonds:

So you see that the molecule has a *bent* shape.

Now you can appreciate that the first step in predicting molecular geometry is to determine the arrangement of the electron domains. Just ignore the lone pairs for a moment and you will have your answer. This method can produce several common shapes. We have seen the two shapes that you can get from a tetrahedral arrangement of domains (if one domain is a lone pair, then you get a trigonal pyramidal shape; and if two domains are lone pairs, then you get a bent shape—these were the last two examples that we just saw). Both of these examples started with a tetrahedral arrangement of electron domains (*four* electron domains).

Now we need to consider the common shapes that you can get when you start with *five* electron domains. Let's go through these shapes methodically.

When we have five electron domains, the arrangement of the electron domains is in a trigonal bipyramidal arrangement. We saw a picture and a drawing of this arrangement:

Now we will consider what shapes we get when one or more of these five domains is a lone pair. In order to do this we will need to introduce some terminology. The three electron domains that are all in the same plane are called *equa-*

torial positions, and the remaining two electron domains are called *axial* positions.

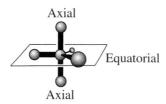

We will need to use this terminology, because it turns out that the orbitals used for equatorial positions are lower in energy than the orbitals used for axial positions. Therefore, if we have to place a lone pair into one of the five orbitals, then we will place it in an equatorial position, rather than an axial position. In fact, if we have two or even three lone pairs, we will place all of the lone pairs in equatorial positions. Now let's see what shapes we can get when we start putting lone pairs into these equatorial positions:

1. If *one* of these five electron domains is occupied by a lone pair, the lone pair will go into an equatorial position to give the following shape:

This is called a *seesaw* shape. To see why, we rotate the image:

2. If *two* of the five electron domains are occupied by lone pairs, the lone pairs will *both* go into equatorial positions:

We call this a ***T-shape***. To see why, just turn your head to the right, and you should see the capital letter T.

3. Finally, if three of the five orbitals are occupied by lone pairs, the lone pairs will all go into equatorial positions, giving a ***linear*** shape:

The three examples above are all of the shapes that are based on a trigonal bipyramidal arrangement of electron domains. With one lone pair, we saw the seesaw shape; with two lone pairs, we saw the T-shape; and with three lone pairs, we saw the linear shape.

Now we are ready to see the last two shapes—the shapes that you can get when you have an octahedral arrangement of electron domains.

1. If we place *one* lone pair into one of the six electron domains of an octahedral geometry, we get the following shape:

I could have placed the lone pair into any one of the six domains, because they are all identical (we don't have axial or equatorial groups in an octahedral arrangement—all six positions are identical). I chose to place the lone pair in the domain pointing straight down, because this is the easiest way for us to see why we call this shape *square pyramidal*. If we draw lines connecting the atoms, we see a pyramid with a square base:

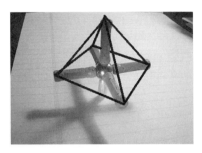

2. In an octahedral arrangement, if two of the electron domains are lone pairs, then the lone pairs will want to be as far apart as possible:

This gives us a shape that we call *square planar*, because the four groups attached to the central atom form a square (contained in one plane):

Now we have seen *all* of the common shapes. Before we do some problems, let's just review all of the information we have seen so far. Do NOT memorize the chart below. Instead, read through the various shapes on the chart, and make sure that it makes sense to you (try to picture it in your mind without looking back).

The first column in the chart indicates the number of electron domains, and this number determines the *arrangement* of the domains. There are five arrangements, and they are shown in the second column below. Then, for each of these five arrangements, you need to consider how the presence of lone pairs can affect the shape (shown in the third and fourth columns). For each shape below, try to picture the shape in your mind. Begin by visualizing the arrangement of the electron domains, and then consider what it looks like when some of these domains contain lone pairs. Once again, DO NOT memorize this chart. Take a few moments and make sure that it makes perfect sense:

# of Domains	Arrangement of Domains	# of Lone Pairs	Shape
2	Linear	0	Linear
3	Trigonal planar	0	Trigonal planar
		1	Bent
4	Tetrahedral	0	Tetrahedral
		1	Trigonal pyramidal
		2	Bent
5	Trigonal bipyramidal	0	Trigonal bipyramidal
		1	Seesaw
		2	T-shape
		3	Linear
6	Octahedral	0	Octahedral
		1	Square pyramidal
		2	Square planar

Now that we have seen all of the common shapes, let's do one more example, and then you can try some problems yourself.

EXERCISE 8.15. Predict the geometry of ClF_5.

Answer: Begin by drawing the Lewis structure. When you apply the method from Chapter 7, you will get the following Lewis structure:

$$F \overset{\displaystyle F}{\underset{\displaystyle F}{\overset{\cdots}{\underset{|}{>Cl<}}}} F$$

Now, just stare at the central atom, and you can determine how many electron domains it has. The central chlorine atom has five bonds and one lone pair, so it has *six* electron domains. Therefore, the arrangement of the domains must be octahedral.

Next, take lone pairs into account. There is only one lone pair, so one of the six domains in the octahedral arrangement has a lone pair. This gives you the square pyramidal shape:

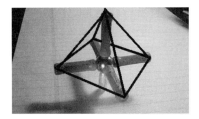

Now let's get some practice. In each of the compounds below, draw the Lewis structure and then predict the shape. In each case, you will need to use the following steps:

1. Draw the Lewis structure, so that you can count the number of electron domains.

2. Use that number to determine the arrangement of the domains.

3. Ignore the lone pairs for a moment, and determine the shape produced by the bonds.

   DO NOT USE THE CHART ABOVE (showing all of the possible shapes). That would defeat the purpose of these exercises. You will not have access to that kind of chart on an exam, so make sure that you can predict the geometry for each of the compounds below *without* the help of the chart.

8.16. PCl_3

8.17. BCl_3

8.18. BeH_2

8.19. $AlBr_4^-$

8.20. CH_4

8.21. ClF_3

8.22. SF_4

8.23. XeF_4

8.24. $NHCl_2$

8.25. $KrBr_2$

8.26. H_3O^+

8.27. IF_5

8.28. $SeCl_4$

8.29. PBr_5

8.30. SCl_6

8.31. XeF_2

8.4 PREDICTING MOLECULAR SHAPE IN COMPOUNDS WITH DOUBLE OR TRIPLE BONDS

So far, we have not seen any examples of compounds containing double bonds or triple bonds. Luckily, the presence of double or triple bonds does not complicate the procedures at all. In fact, we use the exact same procedure as we have in the previous sections. The trick is to just count a double bond as *one* electron domain. Similarly, we just treat a triple bond as *one* electron domain. Before we see an example of how to do this, we should say clearly that a double bond (or triple bond) is NOT one electron domain. However, we treat it that way for purposes of predicting molecular shape, and it turns out that we will get the right answer.

As an example, consider the structure of ethylene:

$$
\begin{array}{cc}
H & H \\
\diagdown & \diagup \\
C & = & C \\
\diagup & \diagdown \\
H & H
\end{array}
$$

The two carbon atoms above have the same story, so let's focus on one of them for now. Let's choose the one on the left side. It is connected to three atoms. Even though one of those connections is a double bond, we nevertheless only count the double bond as one electron domain. So, the carbon atom has a total of three electron domains (for purposes of predicting the shape). When we apply the method that we have seen in this chapter, we will discover that three electron domains will form a trigonal planar arrangement. Since none of the three domains are lone pairs, then the geometry of that carbon atom is trigonal planar.

If we apply the same method to the carbon atom on the right side, we will arrive at the exact same conclusion—it is trigonal planar. There is really only one difference between this and the examples in the previous sections. In all of the examples we have seen before, there was one central atom, and all of the other atoms were connected to that central atom. To determine the shape of the molecule, we just had to determine the geometry of the central atom. But in this case, there is not one central atom. There are actually two (the two carbon atoms). So we must consider the geometry at each of these carbon atoms.

This example points out the difference between atomic geometry and molecular geometry. In the all of the examples in the previous sections, these two terms were essentially the same. By determining the atomic geometry of the central atom, we automatically knew the geometry of the entire molecule. In this case, we need to consider the atomic geometry of two atoms in order to determine the geometry of the entire molecule.

There is a simple way to determine how many atoms you need to look at when you are trying to predict molecular geometry. You just look at the Lewis structure and look for atoms that are connected to more than one other atom. For example, let's take a closer look at the example we just saw (ethylene):

$$
\begin{array}{cc}
H & H \\
\diagdown & \diagup \\
C & = C \\
\diagup & \diagdown \\
H & H
\end{array}
$$

All of the hydrogen atoms are each connected to only one atom (each hydrogen atom is connected to one carbon atom). But each of the carbon atoms is connected to more than one other atom. Each carbon atom is connected to three other atoms:

$$
\begin{array}{cc}
H & H \\
\diagdown & \diagup \\
\textcircled{C} = \textcircled{C} \\
\diagup & \diagdown \\
H & H
\end{array}
$$

So, we will need to determine the atomic geometry around each of these carbon atoms. We just did that, and we discovered that the geometry of each carbon atom is trigonal planar. That means that the entire molecule will be planar:

In the last section of this chapter (when we discuss hybridization states), we will revisit this example so that we can better understand what a double bond is (and why the central bond in the molecule above is actually a double bond). For now, let's just get some more practice predicting geometry when a molecule has a multiple bond:

EXERCISE 8.32.   Predict the molecular geometry of acetylene:

$$H-C\equiv C-H$$

Answer: You'll immediately notice that there is not one central atom here. Each of the carbon atoms is connected to more than one other atom:

$$H-(C\#C)-H$$

So, you will need to determine the geometry of each carbon atom. Start with the one on the left. It has one single bond, one triple bond, and no lone pairs. Count the triple bond as if it is only one electron domain, so the carbon atom has a total of two electron domains. Therefore, the arrangement of the electron domains will be linear.

Apply the same logic to the carbon atom on the right, and you get the same result. So, you can see that both carbon atoms are linear. Therefore, the entire molecule is linear.

Now let's get some practice on compounds that contain double or triple bonds. For each of the following compounds, determine the molecular geometry:

8.33.

$$\ddot{O}$$
$$\|$$
$$H^{\diagdown}C^{\diagup}H$$

8.34. $\ddot{O}=C=\ddot{O}$

8.35. H−C≡N:

8.5 BOND ANGLES

Once you have determined the shape of a molecule, it is generally trivial to predict the bond angles in the molecule. The bond angles are determined by the arrangement of the electron domains. There are not many numbers to memorize—and once your mind can "see" each of the five basic arrangements, then you should not have a hard time memorizing the numbers shown here:

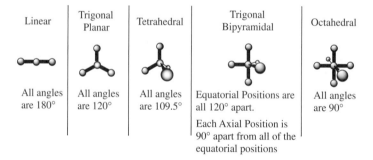

Linear	Trigonal Planar	Tetrahedral	Trigonal Bipyramidal	Octahedral
All angles are 180°	All angles are 120°	All angles are 109.5°	Equatorial Positions are all 120° apart. Each Axial Position is 90° apart from all of the equatorial positions	All angles are 90°

Clearly, the only one that is complicated is the trigonal bipyramidal arrangement, so you should take special note of those angles.

Now that we know the bond angles for each of the five basic arrangements, we must take into account how the presence of lone pairs can very slightly modify the bond angles in these arrangements. For example, consider the structure of H_2O again. We saw that this molecule has four electron domains. These four domains will therefore arrange themselves in a tetrahedral arrangement. In a perfect tetrahedral arrangement, all bond angles will be 109.5°. However, the electron domains in water will not be arranged in a *perfect* tetrahedral arrangement, because the two lone pairs will repel each other more strongly than the bonds repel each other, and this *slightly* changes the bond angles from the perfect tetrahedral arrangement.

That is why the bond angle in water is slightly less than 109.5°. It has been measured as 105°.

So we see that it is possible to make *general* predictions about the bond angles. For example, you should be able to look at a molecule and predict that a particular bond angle will be slightly less than 109.5°. Or perhaps, you will determine that the bond angle will be slightly less than 90° (this would be the case if you have an octahedral arrangement of electron domains, and one of the domains is a lone pair). But, keep in mind that these are only *predictions*. In the end, the only way to verify our predictions is to actually run experiments in a laboratory and determine what the bond angles actually are.

For example, if we try to predict the bond angle of H_2S, we would arrive at the same answer as we did for H_2O. We would conclude that the bond angle should be slightly less than 109.5°. But experiments show that the bond angle is much closer to 90°. This reinforces an important concept: that all we can do is make *predictions*. Usually those predictions will be pretty accurate, but some examples will defy our predictions (especially large atoms, where the repulsion of orbitals is not as strong of a factor). For now, we will just focus on how to make the predictions.

Let's see one more example:

EXERCISE 8.36.   Predict the bond angles in ammonia (NH_3).

Answer: Begin by drawing the Lewis structure:

$$H-\overset{\displaystyle ..}{\underset{\displaystyle |}{N}}-H$$
$$H$$

The nitrogen atom is the central atom, and it has three bonds and one lone pair. Therefore, it has a total of four electron domains. And we have seen that four electron domains can get the farthest apart when they are in a tetrahedral arrangement.

In a tetrahedral arrangement, we expect the bond angles to be 109.5°. But one of the domains is occupied by a lone pair. This lone pair will strongly repel the other three electron domains (the three bonds), and this will cause the bond angles to be slightly less than 109.5° (it has been measured to be 107°).

For each of the following compounds, draw the geometry you would expect, and predict the bond angles:

8.37. PF_5

8.38. PF_3

8.39.   SF$_4$

8.40.   ICl$_5$

8.6   **HYBRIDIZATION STATES**

In the previous sections, we saw that orbitals always want to be as far apart as possible. But we just assumed that they *can* get as far apart as possible.

When we looked at the magnetic toys as an analogy, we saw that the magnets were able to slide along the surface of the metal sphere, and therefore, they were able slide away from each other. But orbitals are ***not*** really able to move around on the surface of the nucleus. They actually have specific positions and shapes, defined by the mathematics that describe them. And if we look at the positions and shapes of atomic orbitals, we will see that the arrangements we discussed don't seem possible.

To see why, let's consider the tetrahedral arrangement (four electron domains). A good example is carbon. Carbon has four valence electrons, so it often forms four single bonds. Therefore, it is common to find a carbon atom with a tetrahedral geometry:

But, we need to ask ourselves how the orbitals are able to assume this arrangement. Carbon is a second-row element:

Second
Row

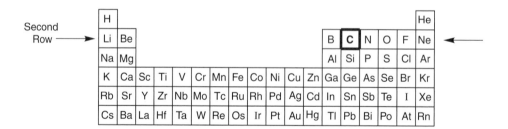

H																	He
Li	Be											B	C	N	O	F	Ne
Na	Mg											Al	Si	P	S	Cl	Ar
K	Ca	Sc	Ti	V	Cr	Mn	Fe	Co	Ni	Cu	Zn	Ga	Ge	As	Se	Br	Kr
Rb	Sr	Y	Zr	Nb	Mo	Tc	Ru	Rh	Pd	Ag	Cd	In	Sn	Sb	Te	I	Xe
Cs	Ba	La	Hf	Ta	W	Re	Os	Ir	Pt	Au	Hg	Tl	Pb	Bi	Po	At	Rn

This means that carbon only has four orbitals available to use: an s orbital and three p orbitals. We talked about this at length in Chapter 7, when we discussed the reason for the octet rule. If we look at the shapes of the s orbital and the three p orbitals, we find that they look like this:

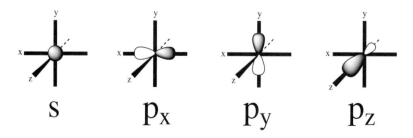

$$s \qquad p_x \qquad p_y \qquad p_z$$

The s orbital is shaped like a sphere. Each of the three p orbitals is aligned along one axis (x, y, or z). Therefore, we designate each p orbital with a subscript to indicate which axis it is on (p_x, p_y, and p_z).

If we try to use these orbitals to form bonds, we will ***not*** get bond angles of 109.5°. We ***won't*** get a tetrahedral shape. Rather, we will see bond angles of 90°, because the p orbitals are separated by 90° (each one is on an axis, as shown above). So how do we use these orbitals to get a tetrahedral arrangement?

Here is the trick: in order to have four orbitals in a tetrahedral arrangement, we will need to have four similar orbitals. Rather than having one s orbital and three p orbitals, we need to have four *equivalent* orbitals (orbitals that have the same shape and the same energy). And that is where hybridization comes in.

Hybridization is really just a mathematical trick. We take the equations that describe the s orbital and p orbitals, and we mathematically average (or hybridize) them to get new orbitals that all look the same. And we call them ***hybridized orbitals***.

Students often have trouble with this concept, so let's use an analogy. Imagine that you are very wealthy and you have four pools in your backyard. One of them is shaped like a triangle and the other three are shaped like pentagons:

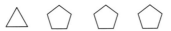

One day, you walk out on to your balcony and you look down at your pools, and it suddenly occurs to you that they are very oddly shaped. You decide that you would rather have four rectangular pools. So, you pull out your magic wand, you chant some magic spell, and presto . . . you now have four rectangular pools; all of them are now exactly the same dimensions:

You have "hybridized" your four oddly shaped pools to get four new, identical pools. This is impressive magic, but there is one severe limitation to this magic: you cannot end up with a different number of pools than you started with. Since you started with *four* pools, you must end up with *four* pools in the end.

When it comes to orbitals, we are doing the same thing. Only, this time, it's not really magic. We are using rigorous mathematical equations that allow us to "average" the orbitals. We are starting with four orbitals (one shaped like a sphere, and the other three shaped like teardrops):

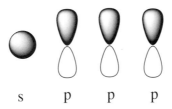

s   p   p   p

For simplicity, we will not draw which axis each p orbital is on (x, y, or z). We are just trying to show that we start with one s orbital and three p orbitals. Now, by applying sophisticated mathematics we get four new, equivalent orbitals. Since we started with *four* orbitals, we MUST end up with *four* orbitals. But now, all four orbitals look the same. They all look like this:

Front
Lobe

Back Lobe

Notice that a hybridized orbital has a front lobe and a back lobe. This is similar to a p orbital, which also has a front lobe and a back lobe, but in p orbitals, the front and back lobes are the same size. In hybridized orbitals, they are different sizes:

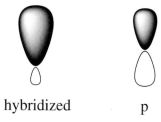

hybridized   p

When we use a hybridized orbital to form a bond or a lone pair, we show that using the front lobe:

Bond   Lone Pair

Remember how we made these hybridized orbitals—we mathematically averaged one s orbital and three p orbitals. Since we started with four orbitals, then we must have a total of four orbitals in the end (in the analogy, we saw that this was the limitation of our magic). Since there are *four equivalent* hybridized orbitals, they will be arranged as far apart from each other as possible, and this explains how we get a tetrahedral arrangement:

For purposes of simplicity, the back lobes of each hybridized orbital were not drawn (that would have made the picture more difficult to see).

Now we need to give these hybridized orbitals a name. Since we made them from one s orbital and three p orbitals, we will call them *sp³ hybridized orbitals*.

If you really want to understand *how* we are able to average orbitals, then you would need to study the mathematics that describes the linear combination of atomic orbitals. This topic is taken up in a typical physical chemistry course. Fortunately, you probably won't need to learn that math during this course, so for now, it will be sufficient to stick with a superficial treatment.

We used hybridization states to describe the tetrahedral shape. What about the trigonal planar shape? How do we get that? Again, we will use hybridized orbitals, but this time, we will do it a bit differently. To see how, let's go back to our pool analogy.

Let's say that you only wanted to average three of your pools, because you wanted to keep one as a pentagon:

So, you wave your magic wand, and you get three rectangular pools and one pentagon:

Notice, once again, that we started with four pools, so we ended up with four pools. And once again, we have just averaged the shapes of the pools. But this time, we did not average *all* of them. We only averaged three of them, leaving one behind, untouched.

We can do the same thing with orbitals. We have one s orbital and three p orbitals to work with, so let's consider what happens if we leave one of the p orbitals untouched. We are averaging one s orbital and two p orbitals. When we average these three orbitals, we get three new orbitals that all look the same (plus the one p orbital that we did not touch, so once again, we have a total of four orbitals). The three new orbitals are called sp² hybridized orbitals, because we formed these orbitals by mathematically combining an s orbital and *two* p orbitals. These sp² hybridized orbitals look very similar in shape to sp³ hybridized orbitals, but they are just a little bit shorter and fatter:

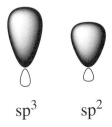

sp^3 $\quad$ sp^2

Now that we know what an sp² hybridized orbital is, the question is: how do three equivalent (sp²) orbitals get as far apart as possible? And here is the answer we are looking for: **_trigonal planar_**. All three orbitals are in one plane:

Once again, the back lobes of each orbital are not shown (to make the picture easier to see). And that is how we explain the trigonal planar shape. But what about that p orbital that we left untouched? What happened to it?

The p orbital is perpendicular to the plane of the sp³ hybridized orbital. For simplicity, we will draw all three sp² hybridized orbital as sticks so that we can clearly see the plane and the p orbital:

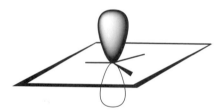

Mathematicians have a special term to describe something that is perpendicular to a plane. The term is **_orthogonal_**. So the p orbital is orthogonal to the plane of the three hybridized orbitals.

So, now we have another kind of **_hybridization state_**. We have already seen a carbon atom that was sp³ hybridized. Now, we know what sp² hybridization is, but why would an atom ever **_need_** to be sp² hybridized?

Let's consider the following compound: AlCl₃. If we draw the Lewis structure of this compound, we get:

$$\begin{array}{c} Cl \\ | \\ Cl^{\diagup Al}\diagdown Cl \end{array}$$

Aluminum is in the third column of the periodic table, and it therefore has three valence electrons. In this example, it is using all three electrons to form bonds (each one in a separate orbital). So, the aluminum atom has no lone pairs in this example. With three bonds and no lone pairs, the aluminum atom is using three orbitals. The fourth orbital is empty (and aluminum therefore does not have an octet). Now we can understand why we left the p orbital untouched, and why we only averaged the s orbital and two of the p orbitals. If we had averaged all four orbitals to get sp^3 hybridized orbitals, we would get a tetrahedral arrangement. But one of those orbitals would be empty, so the remaining three orbitals (the bonds) would not be as far as apart as possible. So, instead, we do not average all of the orbitals. We leave one of the p orbitals untouched and empty, and that gives us three equivalent sp^2 hybridized orbitals. The *empty* p orbital is orthogonal to the plane of the three bonds (formed from the three hybridized orbitals):

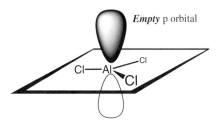

There is one other important situation where we use sp^2 hybridization to explain bond angles, and that is with double bonds. As an example, consider the case of ethylene:

$$
\begin{array}{ccc}
H & & H \\
\diagdown & & \diagup \\
& C=C & \\
\diagup & & \diagdown \\
H & & H
\end{array}
$$

Let's pick one carbon atom and examine it closely. If we look at the carbon on the left side, we see that it is connected to three different atoms: C, H, and H. Now, we know that carbon is in the fourth column of the periodic table, which means it has four valence electrons (with which it can form four bonds). But in this case, the carbon atom is only connected to three different atoms. So, imagine that we use our fancy mathematics and we hybridize the orbitals, using the sp^2 hybridization state. We will get three sp^2 hybridized orbitals, and one p orbital that is orthogonal to the plane of the hybridized orbitals will be left over. Each of the hybridized orbitals gets one electron, which it uses to form a bond. The remaining (fourth) valence electron goes into the orthogonal p orbital:

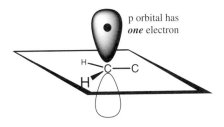

Now, we look at the carbon atom on the right side, and we draw the same conclusion. It is connected to three atoms, so it will be sp^2 hybridized, and the fourth valence electron will go into the orthogonal p orbital:

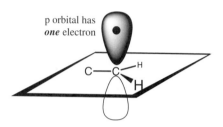

And now we can understand what a double bond is. Those p orbitals are overlapping each other, and that is what gives us the carbon-carbon double bond:

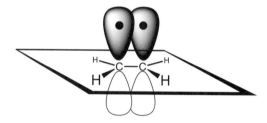

So, if we analyze the double bond closely, we will see that there are two kinds of orbitals overlapping. First, we have the overlapping sp^2 hybridized orbitals:

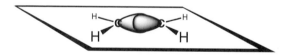

That gives us our first bond between the carbon atoms. Then we have overlapping p orbitals. That gives us our second bond:

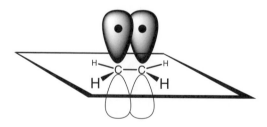

Now we can understand something we said earlier in this chapter. When we learned how to determine the shape of molecules that contain double bonds, we said to treat the double bond as one electron domain, even though it is not. After what we have seen in this section, we can now appreciate that a double bond is really ***not*** only one electron domain after all, but we can also appreciate why we treat it as if it is one domain. In this molecule, if we treat the double bond as one electron domain, then we will count each carbon atom as having only three electron domains. And we said that three electron domains will assume a trigonal planar arrangement. That fits perfectly here. The double bond comes from p orbitals that are orthogonal to the plane of the molecule. When we focus on all of the atoms in the molecule, we see that the molecule is planar. So, by treating a double bond as one electron domain, you will get the correct geometry.

So far, we have seen two hybridization states: sp^3 and sp^2. But there is one other hybridization state that we need to closely consider. Let's go back to our pool analogy.

Let's say that you only wanted to average *two* of your pools, because you wanted to keep two as pentagons:

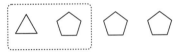

So, you wave your magic wand, and you get two rectangular pools and two pentagons:

Notice, once again, that we started with four pools, so we ended up with four pools. Once again, we essentially averaged the shapes of the pools. But this time, we only averaged *two* of them, leaving two pools behind, untouched.

We can do the same thing with orbitals. We have one s orbital and three p orbitals to work with. When we wanted to get the tetrahedral shape, we averaged *all four* orbitals. When we wanted to get the trigonal planar shape, we averaged the s orbital and *only two* p orbitals, leaving one p orbital untouched. Now we will consider what happens if we average one s orbital and *only one* p orbital. So, we are leaving two of the p orbitals untouched. When we average one s orbital and one p orbital, we get two new orbitals that look similar (plus the two p orbitals that we did not touch, so once again, we have a total of four orbitals). The two new orbitals are again called hybridized orbitals, but we don't call them sp^3 or sp^2 orbitals. We only used one p orbital (and an s orbital) to form these hybridized orbitals. So, we call them *sp hybridized*. They look very similar in shape to sp³ and sp² hybridized orbitals, but they are just a little bit shorter and fatter:

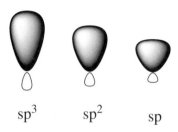

$$sp^3 \qquad sp^2 \qquad sp$$

So, the question is: how do two equivalent (sp) orbitals get as far apart as possible? Answer: in a ***linear*** shape:

Once again, we have not drawn the back lobes, for simplicity. When we don't draw the back lobes, the drawing almost looks like one p orbital (drawn sideways). But be careful, it is not a p orbital. It is *two* sp hybridized orbitals, and we are looking at only the front lobes of each sp orbital. But speaking of p orbitals, what happened to the two p orbitals that we left untouched? Where are they?

To see where they are, let's draw the two sp orbitals as sticks (one going directly to the right, and the other going directly to the left). That way, we can focus on the p orbitals: one p orbital is going straight up and down, and the other p orbital is coming in and out of the page:

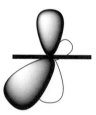

So, now we have another kind of ***hybridization state***. In the sp hybridization state, there are two sp orbitals and two p orbitals. But now we must explore why an atom would ever ***need*** to be sp hybridized?

Consider the following compound: BeH_2. If we draw the Lewis structure of this compound, we get:

H—Be—H

Beryllium is in the second column of the periodic table, and therefore, it has two valence electrons. In this example, it is using both of its electrons to form bonds (each one in a separate orbital). So, the beryllium atom has no lone pairs in this example. With two bonds and no lone pairs, the beryllium atom is using only two orbitals. The remaining two orbitals must be empty (and beryllium therefore does not have an octet). Now we can understand why we left two p orbitals untouched, and why we only averaged the s orbital and one of the p orbitals. If we had averaged all four orbitals to get sp^3 hybridized orbitals, then the two bonds would not be as far apart as possible. Instead, we average an s orbital and one p orbital to give two sp hybridized orbitals. These orbitals get as far apart as possible (linear arrangement), and they are used to form bonds. The other two p orbitals remain *empty*:

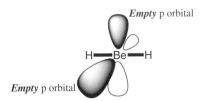

In the drawing above, the sp orbitals are not drawn (they are used to form the bonds to the hydrogen atoms). For simplicity, we have drawn the sp orbitals as straight lines, so that we could focus on the empty p orbitals.

There is one other important situation where we use sp hybridization to explain bond angles, and that is with triple bonds. As an example, consider the case of acetylene:

$$H-C\equiv C-H$$

Let's pick one carbon atom and examine it closely. If we look at the carbon on the left side, we see that it is connected to two different atoms: C and H. Now, we know that carbon is in the fourth column of the periodic table, which means it has four valence electrons (with which it can form four bonds). But in this case, the carbon atom is only connected to *two* different atoms. That means that we only need *two* hybridized orbitals to get the job done. So, imagine that we use our fancy mathematics and we hybridize the orbitals, using the sp hybridization state: we get two sp hybridized orbitals, and we will be left with two untouched, p orbitals. Each of the sp hybridized orbitals gets one electron, which are each used to form a bond (shown as straight lines below). The remaining two valence electrons go into the two p orbitals:

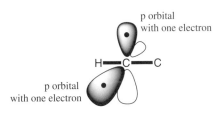

Now, we look at the carbon atom on the right, and we draw the same conclusion. It is connected to two atoms, so it will be sp hybridized, and the remaining two valence electrons will go into the p orbitals, just like in the drawing above.

So, now we can understand what a triple bond is. The p orbitals of each carbon atom are slightly overlapping each other, and that is what gives us the carbon-carbon triple bond:

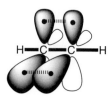

Once again, we can now understand why we treat a triple bond as *one* electron domain, even though it is not. When we learned how to determine the shape of molecules that contain triple bonds, we said to treat the triple bond as one electron domain, even though it is not. After what we have seen in this section, we can now appreciate that a triple bond is really *not* only one electron domain after all, but we can also appreciate why we treat it as if it is one domain. In this molecule, if we treat the triple bond as one electron domain, then we will count each carbon atom as having only two electron domains. And we said that two electron domains will assume a linear arrangement. That fits perfectly here. The triple bond is formed from the two sets of p orbitals overlapping, but they don't really affect the geometry of the molecule. It is still a linear molecule. So, by treating a triple bond as one electron domain, you will get the correct geometry.

In total, we have seen three different hybridization states: sp³, sp², and sp. These are the only possibilities for second-row elements, because second-row elements only have one s orbital and three p orbitals that we can mathematically play with. You can think of these three hybridization states as "packages". Sometimes a salesman will try to sell you something, and he will offer you three different packages to choose from. It is similar here: a second-row element in any molecule will be using one of these three "packages". It will either be sp³ hybridized, sp² hybridized, or sp hybridized.

In our entire discussion of hybridization states, we have focused on second-row elements, but what about third-row elements? They also have access to an s orbital and three p orbitals, but they have access to d orbitals as well. This means that we have more orbitals to play with. And therefore, there are more possible ways to average the orbitals. You can get sp³d hybridized orbitals, and you can get sp³d² hybridized orbitals, etc. In fact, that is how we get the arrangements where you have more than four electron domains. Recall that *five* electron domains will arrange themselves in a trigonal bipyramidal arrangement, and *six* electron domains will arrange themselves in an octahedral arrangement. These arrangements will involve hybridization states that we get when we mathematically average s, p, **and** d orbitals.

Clearly, we need a simple method for looking at an atom and determining its hybridization state. And we are in luck, because there is a very simple method. All we have to do is count up the electron domains. That's it.

If there are *two* electron domains, then the atom is *sp* hybridized (recall the cases of BeH_2 or H—C≡C—H). This should make sense: when an atom has *two* electron domains, then it will need *two* hybridized orbitals, so we mix an s orbital and a p orbital and we get two sp hybridized orbitals.

If there are *three* electron domains, then we will need three hybridized orbitals. So, we will need to average one s orbital and two p orbitals. This will give us three equivalent *sp²* orbitals.

If there are *four* electron domains, then the atom will be *sp³* hybridized. Four domains will require four hybridized orbitals, so we average one s orbital and three p orbitals to get four sp³ orbitals.

If there are *five* electron domains, then we will need five hybridized orbitals. So we will need to average one s orbital, three p orbitals, and one d orbital. This will give us five *sp³d* hybridized orbitals (two axial and three equatorial, as we have seen earlier).

If there are *six* electron domains, then we will need six hybridized orbitals. So we will need to average one s orbital, three p orbitals, and two d orbitals. This will give us six equivalent *sp³d²* hybridized orbitals.

That's all there is to it. You just count up the number of electron domains. That tells you how many hybridized orbitals you will need to have. To summarize:

# of Electron Domains	Hybridization State	Illustration
2	*sp*	
3	*sp²*	
4	*sp³*	
5	*sp³d*	

# of Electron Domains	Hybridization State	Illustration
6	sp^3d^2	

Let's try a problem:

EXERCISE 8.41. Determine the hybridization state of the iodine atom in ICl_3.

Answer: In order to determine the number of electron domains that iodine is using in this example, you will need to know how many single bonds and how many lone pairs the iodine atom has. So, you will need to begin by drawing the Lewis structure:

$$
\begin{array}{c}
Cl \\
| \\
:I: \\
Cl \quad Cl
\end{array}
$$

You can see that the iodine atom in this example has three single bonds and two lone pairs. That means that it has a total of *five* electron domains. So, it will need to use *five* hybridized orbitals. In order to get five hybridized orbitals, you will need to average one s orbital, three p orbitals, and one d orbital. So, the hybridization state will be sp^3d.

In each of the following compounds, identify the hybridization state of the central atom. In each case, you will first need to draw the Lewis structure. Then, use the Lewis structure to count the number of lone pairs and single bonds on the central atom. That will tell you the number of electron domains, which you will use to determine the hybridization state.

8.42. CCl_4 **8.43.** SF_6 **8.44.** SBr_2

8.45. PBr$_3$ **8.46.** PBr$_5$ **8.47.** BCl$_3$

8.48. AlBr$_4^-$ **8.49.** SF$_4$ **8.50.** XeF$_4$

8.51. XeF$_2$ **8.52.** SCl$_6$ **8.53.** ClF$_3$

8.54. IF$_5$ **8.55.** **8.56.** $\ddot{O}{=}C{=}\ddot{O}$

8.57. $NHCl_2$ **8.58.** $SeCl_4$ **8.59.** PBr_5

8.60. H—C≡N   **8.61.** $KrBr_2$ **8.62.** ICl_5

8.63. CH_4 **8.64.** $BeCl_2$ **8.65.** $AlCl_3$

8.66. BF_3

8.67. BrF_5

8.68. NH_4^+

8.69. H_3O^+

CHAPTER 9

EXPONENTS AND LOGARITHMS

Logarithms are used extensively in general chemistry. You will see countless examples throughout this course. Logarithms are central in topics such as acid-base chemistry, equilibrium concentrations, rates of reactions, and many others.

Students are often confused when logarithms appear in equations. Even students with a solid grasp of algebra are not immune from the confusion generated by the word "logarithm". If you feel that you already understand logarithms and if you know how to manipulate them, then perhaps you won't need this chapter. But a little bit of review never hurt anyone.

In this chapter, we will see that logarithms are just "dressed up" exponents. Exponents are really not that hard to understand, and hopefully, you will see that the same is true with logarithms. This chapter is designed to make sure that you are comfortable with exponents, logarithms, and their manipulation. We will begin our discussion with exponents.

9.1 DEALING WITH EXPONENTS

We are familiar with the idea that multiplication is a shortcut for addition. As an example, suppose we wanted to add a bunch of 4's together:

$$4 + 4 + 4 + 4 + 4 + 4 + 4 + 4 + 4 + 4$$

You can count up how many 4's you have, and then add them all together (and you would get 40). But there is a simpler way to make the same statement. We can express it as a multiplication:

$$10 \times 4$$

So, we see that multiplication is a shortcut for addition.

Similarly, exponents are a shortcut for multiplication. As an example, suppose we are multiplying a bunch of numbers together:

$$3 \times 3 \times 3 \times 3 \times 3$$

There is a simpler way to express this, using exponents. Since we are multiplying five 3's, we just say:

$$3^5$$

So, exponents are just a shortcut for multiplication. They allow us to express things in a simpler way. Exponents can be whole numbers, such as 10^1 or 10^2, but they don't have to be whole numbers. For example, consider the following expression:

$$10^{1.5}$$

What does that mean? This often confuses people. We know that 10^1 is just 10. And we know what 10^2 means 10×10, which is just 100. But what is $10^{1.5}$? It might be tempting to say that it is the number that is halfway in between 10^1 and 10^2. In other words, the number that is halfway between 10 and 100, which is 55. But this is NOT what $10^{1.5}$ means. To understand what it actually means, let's consider a special equation, where we have two variables. One variable is an exponent, and the other is not:

$$y = 10^x$$

In this equation, y is not an exponent, but x is an exponent. This means that y and x do not grow at the same rates. For example, if we substitute $x = 1$ into the equation, we find that the value of $y = 10$:

$$y = 10^x = 10^1 = 10$$

If we substitute $x = 2$ into the equation, we find that $y = 100$. If we keep substituting more values of x, we get the following results:

x	$y = 10^x$
1	10
2	100
3	1000
4	10,000
5	100,000

We see that every time we increase the value of x (by just *adding* 1), the effect on y is that it gets *multiplied* by a factor of 10. So, x and y are growing on different scales. If we plot this relationship between x and y (if we generate a plot of the equation $y = 10^x$), we will get this:

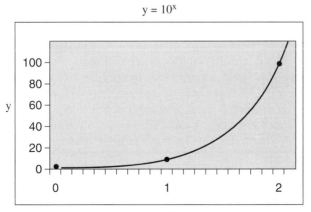

$y = 10^x$

If we connect the dots, we see that we have *a curve* rather than a straight line. The reason for the curve is simple: if we keep increasing the value of x, then y increases at a faster and faster rate. When we increase the value of x from 1 to 2, the effect on y is an increase of 90. When we increase the value of x from 2 to 3, the effect on y is an increase of 900. When we increase the value of x from 3 to 4, the effect on y is an increase of 9000. And that explains why we see a curve that increases at a faster and faster rate. We call this an *exponential curve*.

Now we are ready to use our plot to find out what we mean when we say $10^{1.5}$. Let's locate the halfway point between 10^1 and 10^2 *on the x-axis* of our graph (which corresponds to $10^{1.5}$), and then let's see if we can determine the y value that corresponds to that:

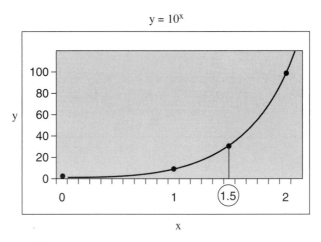

$$y = 10^x$$

The value of $10^{1.5}$ *is* halfway between 10^1 and 10^2 *on the x-axis*, but when we look at the value of y that corresponds to x = 1.5 we see that the value of $10^{1.5}$ is *not* 55 (it is *not* halfway between 10 and 100). The number looks like it is less than 55. In fact, if we use a calculator, we will see that $10^{1.5} = 31.6$. We can say that $10^{1.5}$ *is* halfway between 10^1 and 10^2 if we make the statement carefully: it is halfway between 10^1 and 10^2 *on an exponential scale*.

Similarly, $10^{2.5}$ is *not* halfway between 10^2 and 10^3 (it is not 550). Rather, it is halfway between 10^2 and 10^3 *on an exponential scale.* And that is what we are talking about when we deal with exponents that are not whole numbers.

This graph will be important in the next section when we discuss logarithms. For now, let's make sure that we understand how to use exponents.

Exponents are particularly useful, because they allow us to easily manipulate numbers. There are three simple rules that you need to know:

Rule #1. When we are ***multiplying*** two numbers that are expressed as exponents of the same base, then we just *add* the exponents. As an example, consider the following case:

$$10^{4.5} \times 10^{2.5}$$

Notice that both numbers have a base of 10. In a situation like this, the rule is that we can just add the exponents. So, we get:

$$10^{4.5} \times 10^{2.5} = 10^{(4.5+2.5)} = 10^7$$

This only works when the bases are the same.

Rule #2. When we are **dividing** two numbers that are expressed as exponents of the same base, then we just *subtract* the exponents. As an example, consider the following case:

$$\frac{10^{4.5}}{10^{2.5}}$$

Notice that both numbers have the same base (which is 10). In a situation like this, the rule is that we can just subtract the exponents. So, we get:

$$\frac{10^{4.5}}{10^{2.5}} = 10^{(4.5-2.5)} = 10^2$$

Rule #3. When you have an exponent raised to the power of another exponent, then you *multiply* the exponents. As an example, consider the following case:

$$(10^{23})^2$$

In this case we have a number (10^{23}) that is raised by another exponent (2). So, we just multiply the exponents:

$$(10^{23})^2 = 10^{(23 \times 2)} = 10^{46}$$

Before we move on to logarithms, let's just get some practice with the three rules above.

EXERCISE 9.1. Consider the following expression:

$$\frac{(10^7)^3 \times (10^2)^5}{10^4}$$

Condense this expression using the rules above.

Answer: Begin by noticing that the two terms in the numerator are each expressed as exponents raised to the power of other exponents, so you can condense each of those terms (using Rule #3):

$$\frac{(10^7)^3 \times (10^2)^5}{10^4} = \frac{10^{21} \times 10^{10}}{10^4}$$

Next, you can combine the two terms in the numerator by adding the exponents (using Rule #1):

$$\frac{10^{21} \times 10^{10}}{10^4} = \frac{10^{31}}{10^4}$$

And finally, you can combine the numerator and denominator by subtracting the exponents:

$$\frac{10^{31}}{10^4} = 10^{27}$$

For each of the following problems, condense the expression into one term.

9.2. $10^{23} \times 10^3 = $ _____   **9.3.** $\dfrac{10^{23}}{10^3} = $ _____

9.4. $(10^{23})^3 = $ _____   **9.5.** $10^9 \times 10^4 = $ _____

9.6. $\dfrac{10^{18}}{10^4} = $ _____   **9.7.** $(10^3)^3 = $ _____

9.8. $\dfrac{(10^5)^2 \times (10^3)^4}{10^{10}} = $ _____

You will also encounter negative exponents frequently, so you need to know how to deal with them. You should think of negative exponents like this:

$$10^{-4} = \frac{1}{10^4}$$

Consider the following example:

$$10^6 \times 10^{-3}$$

There are two ways we can condense this. One way is to realize that 10^{-3} is the same as $1/10^3$, so:

$$10^6 \times 10^{-3} = 10^6 \times \frac{1}{10^3} = \frac{10^6}{10^3} = 10^{(6-3)} = 10^3$$

But we could have also condensed this example more quickly using Rule #1:

$$10^6 \times 10^{-3} = 10^{[6+(-3)]} = 10^3$$

We just added the exponents (6 and -3). Both methods gave us the same answer, which confirms that our definition for negative exponents is consistent with our rules for manipulating exponents. Before we move on, let's get some quick practice on a few problems with negative exponents:

9.9. $10^{17} \times 10^{-9} = $ _____   **9.10.** $\dfrac{10^{-23}}{10^3} = $ _____

9.11. $\dfrac{(10^3)^2 \times (10^4)^2}{10^{-8}} = $ _____

9.2 WHAT ARE LOGARITHMS?

Now that we have seen how to interpret and manipulate exponents, we will turn our attention to logarithms. Logarithms are just exponents, dressed up in fancy language. That's right. There is nothing new here.

In the previous section, we saw a plot for the equation $y = 10^x$:

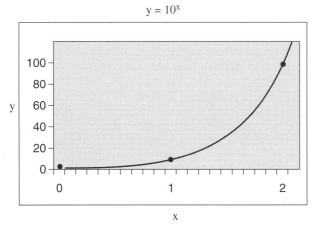

$$y = 10^x$$

The plot shows the relationship between x and y in this special equation. Since x is an exponent and y is not, a slow and steady increase in the value of x will produce a faster and faster increase in the value of y. We have fancy terms that describe the relationship between x and y (as seen on the plot). It is called an *exponential* (or *logarithmic*) relationship. These terms are really just fancy words for a concept that we have already seen in the previous section.

This equation tells us that any number (y) can be expressed in the form 10^x. We just need to know the value of x, such that $10^x = y$. This is why logarithms are so important. Because they allow us to express any number in the format 10^x. But why is that important?

We have already seen how expressing numbers as exponents can be very helpful for simplifying calculations. By capitalizing on the three rules we saw in the previous section, we are able to condense complex calculations into simple ones. So, if we need to perform a series of calculations on many numbers, it can be quicker if we convert all of the numbers into the form 10^x. Mathematicians and scientists have been doing this for the last 400 years, by using logarithms. Before there were calculators, scientists would use a *slide rule* to make calculations. A slide rule is just based on exponential relationships.

But then came the calculator, and that made the slide rule obsolete. So, you might say, "Now that we have calculators, who needs logarithms anymore?" The answer to that question is found in nature. That's right. Nature uses logarithms all of the time. The relationship between chemical properties is often an exponential relationship (and logarithms are just a useful way of expressing exponential relationships). In fact, the term "log" appears regularly in many equations that you will see in this course (equilibrium expressions, rates of reaction, etc.). So even with our wonderful calculators, we can't escape logarithms. We need to know what they are and how to manipulate them.

So what exactly is a logarithm? We said that a logarithm is just another way of expressing exponential relationships. Let's take a closer look, and consider our equation again:

$$y = 10^x$$

This equation tells us that any number (y) can be expressed in the form 10^x. For example, the number 1000 can be expressed as 10 to the power of 3 ($1000 = 10^3$). Similarly, 31.6 can be expressed as 10 to the power of 1.5 (because $31.6 = 10^{1.5}$). We actually saw this last example in the previous section.

Let's say that we start with the number 60. Let's take a look at our plot again (for $y = 10^x$), and find the point on the curve that corresponds to a value of $y = 60$:

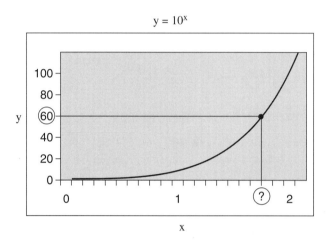

As indicated on this plot, there will be one value of x that corresponds to $y = 60$. In other words, there will be one value of x such that $10^x = 60$. That value of x is called the *logarithm* of 60. Similarly, the *logarithm* of 100 is 2. And the *logarithm* of 1000 is 3.

We can now rewrite the equation above, using our definition:

$$y = 10^x$$

can be rewritten as:

$$\log y = x$$

It is important to be able to switch back and forth between these two equations, so here is a trick that will help you do it quickly: Take the word "log", convert it into the number 10, and then move that 10 to the other side of the equation, like this:

$$\log y = x$$

$$y = 10^x$$

We can go back the other way using the same trick. When we have the expression in terms of $y = 10^x$, we just convert the 10 into the word "log", and move it to the other side of the equation, like this:

$$\boxed{} y = \boxed{10}^{\mathbf{X}}$$

$$\log y = \quad \mathbf{x}$$

This is important, so let's get some practice.

EXERCISE 9.12. Consider the expression below:

$$\log y = 1.900$$

Rewrite this expression in terms of y.

Answer: Using our trick, change the word *log* into a 10, and then move the 10 to the other side of the equation, like this:

$$\boxed{\log} \, y = \boxed{} \, 1.900$$

$$y = 10^{1.900}$$

Now let's do one more example where we convert in the other direction. And this time, we will use a variable:

EXERCISE 9.13. Rewrite the following expression as a logarithm:

$$10^x = 26.7$$

Answer: Using our trick, change the 10 into the word *log*, and then move it to the other side of the equation, like this:

$$\boxed{10}^{\mathbf{X}} = \boxed{} \, 26.7$$

$$\mathbf{x} = \log 26.7$$

If you want to know the value of x, you can just use your calculator to determine the logarithm of 26.7. In fact, you should make sure that you know how to use your calculator to do this. Different calculators work differently, so make sure that you read the instructions, and that you know how to use the logarithm function on your calculator. Your answer should be 1.427.

In each of the following problems, solve for x:

9.14. $10^x = 572.4$ **9.15.** $10^x = 1$

9.16. $10^x = 52$ **9.17.** $10^x = 10,000,000$

9.3 WHAT ARE NATURAL LOGS?

In our discussions so far, we have only dealt with logarithms that use a base of 10:

$$y = 10^x$$

But we could have made the exact same argument for any other base. For example, we could have built a similar argument based on this equation:

$$y = 8.5^x$$

We can manipulate this equation, using a logarithmic system with a base of 8.5, rather than a base of 10. We would just need to specify this in the following way:

$$\log_{8.5} y = x$$

Notice the subscript next to the word log. That subscript tells you the base is 8.5. When the word *log* is used and no base is indicated, it can be assumed that the base is 10, which is why we did not indicate the base in any of our previous examples.

Fortunately, we will only deal with two situations in this course: when the base is 10 (which we have already seen), and when the base is a special number, called *e*. Since we don't deal with any other bases, we will not make any attempts to master logarithms that use other bases. It would take precious time, and you would *not* need that skill in the end (and it could potentially confuse you). So, let's now take a closer look at logarithms that have a base of *e*.

The number *e* is an important constant ($e = 2.7182818\ldots$) that has profound meaning in higher mathematics. For now, we will have to skip all of that. All you need to know is what to do when we use *e* as the base for a logarithm. Consider the following equation:

$$y = e^x$$

This is similar to the equation that we have seen again and again in this chapter ($y = 10^x$). But now, the base is *e* instead of 10. If we plot this equation, we get a plot that looks very similar to the one we saw before, because it is also an exponential curve:

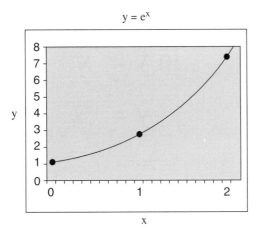

This plot tells us that any number (y) can be expressed in the form e^x. As an example, let's say that we start with the number $y = 6$. Let's take a look at our plot, and find the point on the curve that corresponds to a value of $y = 6$:

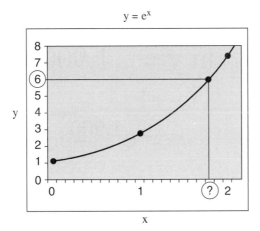

As indicated on this plot, there will be one value of x that corresponds to $y = 6$. In other words, there will be one value of x where $e^x = 6$. That value of x is called the *natural logarithm* of 6, and it is written like this: ln (6).

We can now rewrite the equation above, using our definition:

$$y = e^x$$

can be rewritten as

$$\ln y = x$$

It is important to be able to switch back and forth between these two equations, so we will use the same trick that we used before. But this time we convert the word "ln" into the number e (instead of converting the word "log" into the number 10), and then we move it to the other side of the equation:

$$\ln y = x$$

$$y = e^x$$

This is important, so let's get some practice.

EXERCISE 9.18.   Consider the expression below:

$$\ln y = 1.900$$

Rewrite this expression in terms of y.

Answer: Using our trick, change the word *log* into a 10, and then move the 10 to the other side of the equation, like this:

$$\ln y = 1.900$$

$$y = e^{1.900}$$

PROBLEM 9.19.   Consider the following expression: $\ln y = 2.78$. Solve for y:

Answer: y = _____

PROBLEM 9.20.   Consider the following equation: $\Delta G° = -RT \ln(K)$. Solve this equation in terms of K.

Answer: K = _____

We can go back the other way using the same trick. When we have the expression in terms of $y = e^x$, we just convert the e into the word "ln", and move it to the other side of the equation, like this:

$$\boxed{} y = \boxed{e}\,^X$$

$$\ln y = \quad X$$

EXERCISE 9.21.   Rewrite the following expression as a natural logarithm:

$$e^x = 26.7$$

Answer: Using our trick, change the *e* into the word *ln*, and then move it to the other side of the equation, like this:

$$\boxed{e}\,^X = \boxed{}\,26.7$$

$$X = \ln 26.7$$

Once again, you need to know how to use your calculator to do this. Try to use your calculator to determine ln 26.7. You should get 3.285 as your answer.

In each of the following problems, solve for x:

9.22. $e^x = 275$ **9.23.** $e^x = 1$

9.4 SIGNIFICANT FIGURES
IN LOGARITHMS

Whenever we calculate the logarithm of a number, it is important to report our answer with the correct number of significant figures. We spoke about the importance of significant figures in Chapter 1. For the most part, it was very intuitive. But when it comes to logarithms, it is not so intuitive.

As an example, let's say we need to calculate the logarithm of 65. We type that into our calculator, and we get an answer: 1.81291335664285557. Clearly, not all of those numbers are significant digits. We need to report our answer with the correct number of significant digits. If we look at our original number (65), we see two significant digits. So we might be tempted to report our answer as 1.8. But it is not so simple with logarithms. In order to get this right, we will need to use a bit of terminology.

When reporting a logarithm, the number before the decimal place is called the *characteristic*. The number after the decimal place is called the *mantissa*:

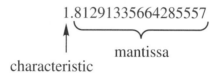

It turns out that we do not count the characteristic as being a significant figure. We only focus on the mantissa. So, in our example, we need to have two significant figures in the mantissa, and we report our answer as 1.81. Notice that our mantissa (the numbers after the decimal place) has two significant figures (1.*81*).

EXERCISE 9.24. Use your calculator to determine the logarithm of 725.4. Report your answer with the correct number of significant figures.

Answer: The original number (725.4) has four significant figures. Therefore, the *mantissa* of your answer must also have four significant figures. So, your answer is: 2.8606

For each of the following problems, use your calculator to make the calculation. Report your answer with the correct number of significant figures.

9.25. log 346.7 = _____

9.26. ln 346.7 = _____

9.27. log 25.006 = _____

9.28. ln 0.0035 = _____

9.29. log 0.00007 = _____

9.30. log 1.00007 = _____

9.5 MANIPULATING LOGARITHMS

Now that you have seen what logarithms are, let's review how to manipulate them. We saw in the previous section that logarithms are really just exponents (with a fancy name). Therefore, we can manipulate logarithms with the same three rules that we saw for manipulating exponents (earlier in this chapter)

Rule #1. When you were learning about exponents, you saw the following rule: When *multiplying* two numbers that are expressed as exponents of the same base, just *add* the exponents:

$$10^a \times 10^b = 10^{(a+b)}$$

Since logarithms are just exponents, you can do a similar manipulation:

$$\log (a \times b) = \log a + \log b$$

Rule #2. When you were learning about exponents, you also saw this rule: When *dividing* two numbers that are expressed as exponents of the same base, just *subtract* the exponents:

$$\frac{10^a}{10^b} = 10^{(a-b)}$$

Since logarithms are just exponents, you can do a similar manipulation:

$$\log (a / b) = \log a - \log b$$

Rule #3. When you were learning about exponents, you also saw this rule: when you have an exponent raised to the power of another exponent, then you *multiply* the exponents:

$$(10^a)^b = 10^{(a \times b)}$$

Since logarithms are just exponents, you can do a similar manipulation:

$$(\log a)^b = b \times \log a$$

EXERCISE 9.31. Use Rules 1–3 to separate the terms in the following expression:

$$\log\!\left(\frac{P_a \times P_b}{P_c}\right)$$

P stands for pressure, and there are three different compounds (compounds a, b, and c) that each have a different pressure. For purposes of rearranging this expression, just treat P_a, P_b, and P_c as three separate variables.

Answer: Use Rule #2 to separate the numerator and denominator:

$$\log\!\left(\frac{P_a \times P_b}{P_c}\right) = \log(P_a \times P_b) - \log(P_c)$$

Then, use Rule #1 to separate the first expression:

$$\log(P_a \times P_b) - \log(P_c) =$$
$$\log(P_a) + \log(P_b) - \log(P_c)$$

Use the three rules for manipulating logarithms to separate the terms in each of the following expressions:

9.32. $\log (V_1/V_2) =$

9.33. $\ln (P_1P_2) =$

9.34. $\log (a \times b \times c)^d$

9.35. $\ln \left(\dfrac{a \times b}{c} \right)^d =$

All of the rules (that you just practiced) work in the opposite direction as well. For example, the reverse of Rule #1 is: $\log (a) + \log (b) = \log (a \times b)$. This allows us to combine logarithmic terms, rather than separating them. Here are the three rules, stated in reverse order:

Rule #1. $\log a + \log b = \log (a \times b)$

Rule #2. $\log a - \log b = \log (a / b)$

Rule #3. $b \times \log a = (\log a)^b$

Now let's get some practice combining logarithmic terms:

EXERCISE 9.36.   Combine the following logarithmic terms into one expression, using the rules above:

$$\log P_1 - \log P_2 - \log P_3$$

Answer: The expression tells you that you are subtracting the second and third logarithmic terms, so use Rule #2 and you get:

$$\log P_1 - \log P_2 - \log P_3 = \log\left(\frac{P_1}{P_2 \times P_3}\right)$$

In each of the following problems, combine the logarithmic terms into one expression:

9.37. $\ln (P_2) - \ln (P_1) =$

9.38. $\ln (P_2) + \ln (P_1) =$

9.39. $\log (K_1) - \log (K_2) =$

9.40. $\log (a) - \log (b) + \log (c) =$

A MATHEMATICAL APPROACH TO BALANCING REACTIONS

Most reactions can be balanced very easily. But once in a while, you might encounter a reaction that seems too difficult to balance. It can be very frustrating because you have to go back and forth *so many times*, and you keep changing your coefficients again and again. Let's see an example like this.

On a separate piece of paper, try to balance the reaction below. This problem is designed to give you a headache. That is the whole point. Try to do it, so you can see what we mean by a *frustrating* problem. Then, when you have convinced yourself that this is indeed frustrating, we will learn a new method for balancing reactions that will make this problem more approachable.

$$CaSO_4 + CH_4 + CO_2 \longrightarrow CaCO_3 + S + H_2O$$

Try to work on this problem for a few minutes. If you are able to solve this problem, then you are very good at balancing reactions, and perhaps you don't need any other tricks. But the majority of us will find this problem to be very difficult. In trying to solve it, we end up going back and forth so many times. Each time we change a coefficient, we have to go back to the other side to compensate for the change we just made. And it just takes too long to balance the reaction.

There is another way to balance reactions, but it is not taught in any of the textbooks. Think about what we are trying to calculate when we balance a reaction. We are looking for *numbers*. We are trying to determine the coefficients necessary to balance the reaction. In the end, our answer is *just a few numbers*. So there is an obvious question: is there a way to construct some kind of formula or equation that will spit out the numbers we want? The answer is: yes, there is. The type of mathematics that we need is algebra.

If you feel that your algebra skills are a bit weak, then perhaps this method will be more confusing than helpful. But if your algebra skills are strong, then you might feel that the following approach can help you when you are in a desperate situation where you can't figure out how to balance the reaction.

This mathematical approach has one major advantage and one major *dis*advantage. The advantage is that it will allow you to easily and methodically balance even the most difficult problems. And once you get practice with this approach, you will find that you can balance most reactions in about 2 minutes. The disadvantage is that it will take you 2 minutes to balance any reaction; even a simple one. Most problems are so easy that you can solve them in jut a few seconds using the regular approach (found in Chapter 3). So, in most situations, you will not want to waste 2 minutes using the mathematical approach. This approach is just to help you when you get stuck. Here is how it works.

Let's start out with a simple example. Then, we will come back and apply the method to more "difficult" reactions, like the one cited earlier.

Consider the following reaction:

$$C_4H_{10} \; (l) + O_2 \; (g) \longrightarrow CO_2 \; (g) + H_2O \; (l)$$

In order to balance this reaction, we essentially need to calculate four coefficients:

$$\underline{a} \; C_4H_{10} \; (l) + \underline{b} \; O_2 \; (g) \longrightarrow \underline{c} \; CO_2 \; (g) + \underline{d} \; H_2O \; (l)$$

Using the language of algebra, we will use the term *variables* to describe the four coefficients. There are four variables that we need to calculate: *a*, *b*, *c*, and *d*.

In order to see how we can construct equations that will help us solve for these variables, think about what it means to balance a reaction. It means that the number and type of atoms on the left side must be the same as the number and type of atoms on the right side. Let's make a list of all of the elements present in the reaction (just like we did at the beginning of every problem in Chapter 3). In this problem, we have: C, H, and O. Therefore, in order to balance this reaction, we must make sure that:

1. there are exactly the same number of *carbon atoms* on both sides of the reaction,

2. there are exactly the same number of *oxygen atoms* on both sides of the reaction,

3. there are exactly the same number of *hydrogen atoms* on both sides of the reaction.

These three statements can be expressed as three separate equations. Let's start with the first equation, for carbon.

Carbon appears in one compound on the left side of the equation (*a* C_4H_{10}). In this compound, carbon appears four times, because the subscript for carbon is 4. So, the number of carbon atoms on the left side of the equation will be equal to *4a*. If the coefficient *a* turns out to be 1 in the balanced reaction, then there will be four carbon atoms on the left side. If *a* turns out to be 2 in the balanced reaction, then there will be eight carbon atoms on the left side. Now let's look at the right side of the reaction. Carbon appears in one compound on the right side (*c* CO_2). In this compound, carbon appears only once, because there is no subscript next to the carbon in CO_2. So, the number of carbon atoms on the right side of the reaction will just be *c*. Now we are ready to form an equation. In order to balance the reaction, the number of carbon atoms on the left side (*4a*) must be equal to the number of carbon atoms on the right side (*c*). So, in order for the reaction to be balanced, we can say that:

$$4a = c$$

This is our first equation, and it tells us the relationship between the coefficients a and c, but it does not solve the problem for us completely. To solve the problem, we need the other equations. If we do the same procedure for oxygen, we will get our second equation. Then we can do the same thing for hydrogen again to get our third equation. Now let's determine what our second equation is (focusing on oxygen). Let's look at the reaction again:

$$\underline{a}\ C_4H_{10}\ (l) + \underline{b}\ O_2\ (g) \longrightarrow \underline{c}\ CO_2\ (g) + \underline{d}\ H_2O\ (l)$$

Oxygen appears in one compound on the left side of the equation ($\underline{b}\ O_2$). In this compound, oxygen appears twice. So, the number of oxygen atoms on the left side of the reaction will be $2b$. Now let's look at the right side. Oxygen appears in two compounds on the right side. In the first compound ($\underline{c}\ CO_2$), oxygen appears two times, and in the second compound ($\underline{d}\ H_2O$), oxygen appears only once. So, the number of oxygen atoms on the right side of the reaction will be $2c + d$. Now we are ready to form an equation. In order to balance the reaction, the number of oxygen atoms on the left side ($2b$) must be equal to the number of oxygen atoms on the right side ($2c + d$). So, in order for the reaction to be balanced, we can say that:

$$2b = 2c + d$$

And this is our second equation.

Finally, we do the same analysis for hydrogen atoms to get our third equation. Hydrogen appears in one compound on the left side ($\underline{a}\ C_4H_{10}$). Since it appears ten times in this compound, the number of hydrogen atoms on the left side of the reaction will be $10a$. On the right side of the reaction, hydrogen appears in one compound ($\underline{d}\ H_2O$). Hydrogen appears twice in this compound, so the number of hydrogen atoms on the right side of the reaction will be $2d$. In order to balance the reaction, the number of hydrogen atoms on the left side ($10a$) must be equal to the number of hydrogen atoms on the right side ($2d$). So, in order for the reaction to be balanced, we can say that:

$$10a = 2d$$

And that is our third equation. Now we can summarize the information we have constructed so far:

$$4a = c$$
$$2b = 2c + d$$
$$10a = 2d$$

These three equations describe the relationships between the coefficients. And we can use these relationships to now calculate the values of these variables. We begin by assigning the number 1 to one of the variables. It doesn't matter which one we choose, but when we look at the equations above, it would make sense to choose a, because we would then immediately know the values of c and *d* (look at the first and third equations above). If we say that $a = 1$ (which acts as our fourth equation), then the other three equations tell us that c will be $4(1) = 4$, and d will be $\frac{10(1)}{2} = 5$. So, now we know a, c, and d. All we need is b and then we will be done. And we can use the second equation to calculate b, because we have the values of c and d. Plugging in these values, we get that $2b = 2(4) + (5) = 13$. So, $b = {}^{13}\!/_{2}$. Now we have finished our calculations, and we can plug our answers into the reaction:

$$C_4H_{10}\ (l) + {}^{13}\!/_{2}\ O_2\ (g) \longrightarrow 4\ CO_2\ (g) + 5\ H_2O\ (l)$$

This reaction is now balanced, BUT we are not using whole numbers as coefficients. As we mentioned earlier, it is always preferable to use whole number coefficients. In order to get rid of the fraction above (before the O_2 on the left side), we multiply everything by 2, and we get our answer:

$$2\ C_4H_{10}\ (l) + 13\ O_2\ (g) \longrightarrow 8\ CO_2\ (g) + 10\ H_2O\ (l)$$

And we are done. Before walking away from the problem, it is always a good idea to double-check to make sure that you actually balanced the reaction properly. You can do this by looking at the each of the elements (on both sides of the reaction) and make sure that you have the same number on either side. In the reaction above, you have 8 carbon atoms on the left and 8 on the right, so C is balanced. You have 20 hydrogen atoms on the left and 20 on the right, so H is balanced. Finally, you have 26 oxygen atoms on the left and 26 on the right, so O is balanced as well. Now we know that we didn't make any mistakes along the way.

It seems like a lot of work. But once you get enough practice using this method, you will find that you will be able to use this method to solve almost any problem in about 2 minutes. In some cases, this could be very helpful. So, we need to get some practice.

But first, we will do another example where we will need to manipulate the equations after we have written them down. To illustrate this, consider the following example:

$$NO_2\ (g) + H_2O\ (l) \longrightarrow HNO_3\ (aq) + NO\ (g)$$

In order to balance this reaction, we essentially need to calculate the following four coefficients:

$$\underline{a}\ NO_2\ (g) + \underline{b}\ H_2O\ (l) \longrightarrow \underline{c}\ HNO_3\ (aq) + \underline{d}\ NO\ (g)$$

We need to calculate: a, b, c, and d.

There are three elements appearing on both sides of the reaction: N, O, and H. So, we will be able to write three equations:

to balance the nitrogen atoms, it must be that:	$a = c + d$
to balance the oxygen atoms, it must be that:	$2a + b = 3c + d$
to balance the hydrogen atoms, it must be that:	$2b = c$

Make sure that these three equations make sense to you, using the method you saw in the previous example.

We now have three equations, and we must start by choosing one variable to be equal to 1. When looking at the last equation above ($2b = c$), it makes sense to choose either b or c to be equal to 1. If we choose that $b = 1$ (this is our fourth equation), then the last equation tells us that $c = 2(1) = 2$. Now we know values for b and c, but we still need to calculate values for a and d. So we look at our other two equations (the first and second equations). Let's rewrite these two equations, substituting the values that we already know ($b = 1$ and $c = 2$):

$$a = 2 + d$$
$$2a + 1 = 3(2) + d$$

Now we have two equations and two unknowns. This is a typical algebra problem. We solve it like this: The first equation tells us that $a = 2 + d$. So we substitute that value of a into the second equation, and we get:

$$2(2 + d) + 1 = 3(2) + d$$

Since this equation only has one variable, we can solve for that variable, and we find that:

$$d = 1$$

And if $d = 1$, then we can solve for a, because $a = 2 + d = 2 + 1 = 3$. Now we have solved the values for all four variables, and we can plug them into our reaction:

$$3\ NO_2\ (g) + H_2O\ (l) \longrightarrow 2\ HNO_3\ (aq) + NO\ (g)$$

Always double-check your answer by looking at each element to make sure it is balanced. When we do this for the reaction above, we see that the reaction is balanced. And in this example we don't have to multiply by any number at the end, because all of the coefficients are already whole numbers. So, we are done.

Now that we have seen how the mathematical approach works, let's use this method to solve our "difficult" problem:

EXERCISE A.1. Using the mathematical approach that we have seen in this section, balance the following reaction:

$$CaSO_4 + CH_4 + CO_2 \longrightarrow CaCO_3 + S + H_2O$$

Answer: Begin by rewriting your reaction with coefficients:

$$a\ CaSO_4 + b\ CH_4 + c\ CO_2 \longrightarrow d\ CaCO_3 + e\ S + f\ H_2O$$

In this example, there are six variables that have to be calculated (a, b, c, d, e, and f).

Now, construct your equations. There are five elements present in this reaction (Ca, S, O, C, and H). So, you will have five equations. Try to write out all five equations before looking at them below, and then check to see if you got them correct:

Ca: $a = d$

S: $a = e$

O: $4a + 2c = 3d + f$

C: $b + c = d$

H: $4b = 2f$

Next, choose one variable to be equal to 1 (and this will act as your sixth equation). If you look at the five equations, the logical choice is a. If you set $a = 1$, then $d = 1$, and $e = 1$. But you still need to calculate b, c, and f. To do this, use the last three equations. There are three equations and three unknowns. This requires the same kind of algebra that you used in the previous problem. When you substitute the values you know (a, d, and e) into the last three equations, you get:

$$4 + 2c = 3 + f$$
$$b + c = 1$$
$$4b = 2f$$

You can manipulate the last two equations so that you will have both c and f in terms of b:

$$c = 1 - b$$
$$f = 2b$$

Then, substitute these values into in the first of the three equations above:

$$4 + 2c = 3 + f$$
$$4 + 2(1 - b) = 3 + (2b)$$
$$4 + 2 - 2b = 3 + 2b$$
$$3 = 4b$$
$$b = \tfrac{3}{4}$$

When you substitute this value of b into the equations above, you will get the values of c and f:

$$c = 1 - b = 1 - \tfrac{3}{4} = \tfrac{1}{4}$$
$$f = 2b = 2(\tfrac{3}{4}) = 6/4 = 3/2$$

Now you have calculated all of your values. You have: $a = 1$, $b = \tfrac{3}{4}$, $c = \tfrac{1}{4}$, $d = 1$, $e = 1$, and $f = \tfrac{3}{2}$. When you put these values into the reaction, you get:

$$CaSO_4 + \tfrac{3}{4}\, CH_4 + \tfrac{1}{4}\, CO_2 \longrightarrow CaCO_3 + S + \tfrac{3}{2}\, H_2O$$

Now you have a balanced reaction, but many of the coefficients are fractions. To get rid of the fractions, multiply by 4, and you get your answer:

$$4\, CaSO_4 + 3\, CH_4 + CO_2 \longrightarrow 4\, CaCO_3 + 4\, S + 6\, H_2O$$

Finally, double-check your answer to make sure each of the elements is balanced. There are 4 calcium atoms on each side of the reaction, 4 sulfur atoms on each side, 18 oxygen atoms on each side, 4 carbon atoms on each side, and 12 hydrogen atoms on each side. So, the reaction is properly balanced.

In general, when you use this approach, you will use an entire sheet of paper. And your piece of paper should be logically structured in the form of a template, so that you always know the next step that you need to do. That way, you can fly through the process. Your paper should be structured according to the following template:

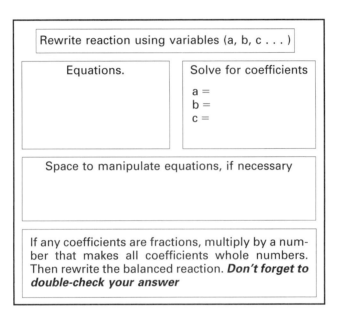

By following this template, it will make the process go faster, and you will get used to using this approach.

There is one final piece of advice that I must give you before you try the problems below. After writing out your equations, you will often find yourself trying to do the simple arithmetic *in your head*. For example, you might see a problem where you determine that $a = 1$. Then, you see that one of your equations says that $a = 2d$. So, you do the math in your head, and you write down that $d = 2$. **WRONG!** ***Don't do it in your head.*** Instead, actually write it out:

$$a = 2d$$
$$(1) = 2d$$
$$\tfrac{1}{2} = d$$

So, we see that $d = \tfrac{1}{2}$ in this case; not 2. That mistake is so easy to make, and it will mess up all of your work, especially if you will then use the value of d to calculate the remaining coefficients. The best way to avoid this kind of mistake is to always write everything out—*never do the simple math in your head*. That's when silly mistakes happen.

Now, let's get some practice. Using a blank piece of paper (and following the template above), try to balance each of the following reactions using the mathematical approach:

PROBLEM A.2. $Al(OH)_3 + HClO_4 \longrightarrow Al(ClO_4)_3 + H_2O$

PROBLEM A.3. $Cu(NO_3)_2 \ (s) + KOH \ (aq) \longrightarrow Cu(OH)_2 \ (s) + KNO_3 \ (aq)$

PROBLEM A.4. $C_2H_2Cl_4 + Ca(OH)_2 \longrightarrow C_2HCl_3 + Ca \ Cl_2 + H_2O$

We said before that most reactions are very easy to balance using the regular approach to balancing reactions (shown in Chapter 3), and it only takes a few seconds. In these situations, you would not want to waste 2 minutes using the mathematical approach. So, for any problem, you should first try to balance it the regular way. If you can solve it quickly, then you're done. If you find that you are stuck, then you can rely on the mathematical approach as a backup plan.

Try to balance the following reactions. Some are designed to be simple, and you will want to use the regular approach to solve them quickly. But some of the following reactions are a bit more complex, and you might find that you are stuck. If you do get stuck, then you can try the mathematical approach to solve them. You decide which method you want to use, but you should always try the regular approach first before moving on to the mathematical approach. Use a separate sheet of paper for each problem and apply the template.

PROBLEM A.5. $NaCl \ (aq) + H_2O \ (l) \longrightarrow NaClO_3 \ (aq) + H_2 \ (g)$

PROBLEM A.6. $Al \ (s) + NH_4ClO_4 \ (s) \longrightarrow Al_2O_3 \ (s) + AlCl_3 \ (s) + NO$
$(g) + H_2O \ (l)$

PROBLEM A.7. $C_3H_8 \ (g) + O_2 \ (g) \longrightarrow CO_2 \ (g) + H_2O \ (g)$

PROBLEM A.8. $CaSiO_3 \ (s) + HF \ (g) \longrightarrow SiF_4 \ (g) + CaF_2 \ (s) + H_2O \ (l)$

PROBLEM A.9. $C_3H_6 \ (g) + NH_3 \ (g) + O_2 \ (g) \longrightarrow C_3H_3N \ (g) + H_2O \ (l)$

ANSWERS

CHAPTER 1

1.2. 0.0713200

1.3. 7843000

1.4. 1.4800

1.5. 100

1.6. 100.0

1.7. 894.003

1.8. 89400

1.9. 0.03000

1.10. 74.000

1.11. 3

1.12. 1

1.13. 3

1.14. 3

1.15. 2

1.16. 1

1.17. 1

1.18. 2

1.19. 2

1.20. 2

1.21. 2

1.22. 1

1.23. 2

1.24. 0

1.25. 35

1.26. 24

1.27. 18

1.28. 18

1.29. 18

1.30. 20

1.31. 46

1.32. 46

1.34. 25

1.35. 0.006322

1.36. 60.6

1.37. 362.5

1.38. 368

1.39. 370

1.40. 0.000204

1.41. 5.0

1.42. 46.3

1.43. 7.3

1.45. 1.43527×10^3

1.46. 2.431×10^2

1.47. 3.801×10^3

1.48. 2.7430×10^2

1.50. 2.43×10^2

1.51. 3.72567×10^5

1.52. 2.8942×10^4

1.53. 4.89×10^3

1.54. 2.4×10^2

1.55. 3.7×10^3

1.57. 5.67×10^{-4}

1.58. 2.4007×10^3

1.59. 3.7024×10^2

1.60. 5.64×10^{-6}

1.61. 4.32×10^{-3}

1.62. 3.840×10^{-2}

1.64. 2480

1.65. 0.00464

1.66. 0.07924

1.67. 792.4

1.68. 0.0008456

1.69. 384000

1.71. 2.01×10^{10}

1.72. 3.6×10^{-33}

1.73. 2.53×10^5

1.74. 5.2×10^{-8}

1.75. 3.2×10^{12}

1.77. 2.94×10^{-4}

1.78. 1.7×10^{10}

1.79. 8.56×10^{11}

1.80. 1.3×10^2

1.81. 2.7×10^3

1.82. 3.4×10^4 m

1.83. 2×10^{-6} s

1.84. 4.82 m

1.85. 2.45×10^2 dm

1.86. 3.72×10^{-8} m

1.87. 1.487×10^3 g

1.88. 2.46×10^{-3} g
1.89. 3.7×10^{-3} s
1.90. g $\times$ cm^{-3}
1.91. kg $\times$ m $\times$ s^{-2}
1.92. mol $\times$ m^{-3}
1.93. g $\times$ cm^2 s^{-2}
1.94. g $\times$ cm^{-3}
1.95. kg
1.96. g $\times$ m^{-3}
1.97. cm $\times$ s^{-1}
1.98. cm
1.99. kg $\times$ mol^{-1}
1.101. 69.49 kg
1.102. 4.4 lbs
1.103. 139 cm
1.105. 1.88×10^3 yards
1.107. 1.2×10^7 cm^3
1.109. 765.9 miles/h
1.110. 0.59 km/min
1.112. 2.9 g $\times$ cm^{-3}
1.114. 6.730×10^3 g
1.115. 1.75×10^3 mL (or 1.75 L)

CHAPTER 2
2.2. CH_3
2.3. C_3H_8
2.4. CO_2
2.5. H_2SO_4
2.6. CH_2O
2.7. HgI
2.8. Al_2O_3
2.9. C_5H_5N
2.11. 1.6×10^{23} atoms
2.12. 1.58×10^{24} atoms
2.13. 3.3×10^{24} atoms
2.15. 253.33 u
2.16. 153.41 u
2.17. 187.77 u
2.19. 20.07 g/mol
2.20. 44.10 g/mol
2.21. 44.01 g/mol
2.22. 82.08 g/mol
2.23. 180.16 g/mol
2.24. 654.98 g/mol
2.25. 101.96 g/mol
2.26. 79.10 g/mol
2.28. 160 g
2.29. 3.73 moles
2.30. 1.42×10^3 moles

2.31. 31.6 moles
2.32. 7.05×10^2 g
2.34. 143.32 g/mol; 9.085 g of Ag
2.35. 100.09 g/mol; 0.88 g of Ca
2.36. 1.2 g of Cu
2.37. 6.28 g of Ag
2.38. molar mass; moles
2.39. molar mass; grams
2.41. 87.6 cm^3
2.42. 97 cm^3
2.44. 0.18 moles
2.45. 0.33 moles
2.46. 0.74 moles
2.48. 36.11%
2.49. 63.89%
2.50. 67.10%
2.51. 54.53%
2.52. 71.08%
2.54. C_2H_6O
2.55. SO_3
2.56. PH_3
2.57. CH_2O
2.58. SO_2
2.60. C_6H_6
2.61. HNO_2
2.62. HIO_3

CHAPTER 3
3.2. $2\ Na + H_2 + 2\ C + 3\ O_2 \longrightarrow$
$2\ NaHCO_3$
3.3. $4\ Al + 3\ O_2 \longrightarrow 2\ Al_2O_3$
3.4. $H_2 + N_2 + 3\ O_2 \longrightarrow 2\ HNO_3$
3.5. $F_2 + H_2 \longrightarrow 2\ HF$
3.6. $2\ Cs + Cl_2 \longrightarrow 2\ CsCl$
3.7. $16\ Cu + S_8 \longrightarrow 8\ Cu_2S$
3.9. $Cr_2O_3 + 3\ Mg \longrightarrow$
$2\ Cr + 3\ MgO$
3.10. $CH_4 + H_2O \longrightarrow CO + 3\ H_2$
3.11. $2\ H_2CO + O_2 \longrightarrow$
$2\ CO + 2\ H_2O$
3.12. $2\ NO + O_2 \longrightarrow 2\ NO_2$
3.13. $2\ NO_2 + F_2 \longrightarrow 2\ NO_2F$
3.14. $4\ CH_3NH_2 + 9\ O_2 \longrightarrow$
$4\ CO_2 + 10\ H_2O + 2\ N_2$
3.15. $3\ Fe + 4\ H_2O \longrightarrow$
$Fe_3O_4 + 4\ H_2$
3.17. $P_2O_5 + 3\ H_2O \longrightarrow 2\ H_3PO_4$

3.18. $NH_4NO_3 \longrightarrow N_2O + 2 H_2O$

3.19. $B_2O_3 + 6 NaOH \longrightarrow$
$2 Na_3BO_3 + 3 H_2O$

3.20. $BCl_3 + 3 H_2O \longrightarrow$
$H_3BO_3 + 3 HCl$

3.21. $(NH_2)_2CO + H_2O \longrightarrow$
$CO_2 + 2 NH_3$

3.22. $C_7H_8O_2 + 8 O_2 \longrightarrow$
$7 CO_2 + 4 H_2O$

3.23. $As_4S_6 + 9 O_2 \longrightarrow$
$As_4O_6 + 6 SO_2$

3.24. $6 S_2Cl_2 + 16 NH_3 \longrightarrow$
$S_4N_4 + S_8 + 12 NH_4Cl$

3.25. $2 Ca_3(PO_4)_2 + 6 SiO_2 + 10 C$
$\longrightarrow P_4 + 6 CaSiO_3 + 10 CO$

3.26. $4 CaSO_4 + 3 CH_4 + CO_2$
$\longrightarrow 4 CaCO_3 + 4 S + 6 H_2O$

3.27. $NaCl\ (aq) + 3 H_2O\ (l) \longrightarrow$
$NaClO_3\ (aq) + 3 H_2\ (g)$

3.28. $SO_3\ (g) + H_2O\ (l) \longrightarrow$
$H_2SO_4\ (l)$

3.29. $2 SO_2\ (g) + O_2\ (g) \longrightarrow$
$2 SO_3\ (g)$

3.30. $N_2H_4\ (l) + 3 O_2\ (g) \longrightarrow$
$2 NO_2\ (g) + 2 H_2O\ (l)$

3.31. $(CN)_2\ (g) + 4 H_2O\ (l) \longrightarrow$
$H_2C_2O_4\ (aq) + 2 NH_3\ (g)$

3.33. 3 moles

3.34. 6 moles

3.35. 2 moles

3.36. 4 moles

3.37. 2 moles

3.39. 0.2636 mol

3.40. 0.5551 mol

3.41. 0.5284 mol

3.42. 0.1101 mol

3.44. 24.41 g

3.45. 26.91 g

3.46. 67.95 g

3.48. 21.69 g

3.49. 1.96 g

3.50. 17.0 g

3.52. 1.04 g

3.53. 7.59 g

3.54. 18.2 g

3.55. 44.7 g

3.57. 94.39%

3.58. 93.0%

3.59. 88.5%

3.60. 94.61%

CHAPTER 4

4.2. 1.324 atm

4.3. 1.654 atm

4.4. 5.0×10^{-2} atm

4.5. 2.50 atm

4.6. 1.18 atm

4.7. 0.609 atm

4.8. 1.49 atm

4.9. 0.138 atm

4.11. 1.342 L

4.12. 1.37 L

4.13. 0.896 L

4.14. 47.3 L

4.15. 21 L

4.16. 0.212 L

4.17. 0.356 L

4.18. 5.45×10^{-2} L

4.20. 2.15 mol

4.21. 1.86 mol

4.22. 2.7×10^{-3} mol

4.23. 0.75 mol

4.24. 7.09 mol

4.26. 373 K

4.27. 273 K

4.28. 298.9 K

4.29. 310 K

4.30. 258 K

4.31. 373 K

4.33. $n = 7.8 \times 10^{-2}$ mol
$V = 2.7$ L
$P = 2.12$ atm

4.34. $n = 0.17$ mol
$V = 3.7$ L
$T = 293$ K

4.35. $n = 0.831$ mol
$V = 5.00 \times 10^{-2}$ L
$P = 1$ atm

4.37. 2.45×10^{-2} atm

4.38. 7.21×10^3 K

4.39. 17.6 L

4.40. 31.3 mol

4.43. First, use V and density to calculate mass. Then, use mass and molar mass to calculate n. Then, use P, V, and n to calculate T.

4.44. First, use P, n, and T to calculate V. Then, use V and density to calculate mass. Then, use mass and n to calculate molar mass.
4.45. Use mass and molar mass to calculate n. Then, use P, n, and T to calculate V. Then, use V and mass to calculate density.
4.46. Use mass and density to calculate V. Then, use mass and molar mass to calculate n. Then, use V, n, and T to calculate P.
4.49. 1.35 g/L
4.50. 105 K
4.51. 0.939 g/L
4.52. 5.69 atm
4.54. 1.86 g/L
4.55. 1.29 g/L
4.56. 0.824 atm
4.57. 426 K
4.58. 16 g/mol
4.60. 7.13 g/L
4.61. 0.211 atm
4.62. 2.1×10^3 K
4.65. 1.05×10^3 K
4.66. 4.56 L
4.67. 259 K
4.69. 138 g CO_2
4.70. 6.46 L
4.71. 29.3 L
4.73. 5.23×10^4 mL
4.74. 30.9 g NH_3
4.75. 23.2 L
4.76. 228 g CO_2

CHAPTER 5
5.2. 8.89 cal
5.3. 25.00 kcal
5.4. 1.1×10^2 cal
5.5. 0.1523 kcal
5.6. 123 J
5.7. 452.7 kJ
5.8. 1.8×10^2 J
5.9. 3.733 kJ
5.11. -1.2 kJ
5.12. -1.2 kJ
5.13. The system must do 0.27 kJ of work; in other words, w = -0.27 kJ.
5.14. q = 0.81 kJ
5.16. q = $+0.25$ kJ. Therefore, the reaction is endothermic.

5.17. q = -0.18 kJ. Therefore, the reaction is exothermic.
5.20. $\Delta H° = 21$ kJ
5.21. $\Delta H° = -156$ kJ
5.23. $\Delta H° = 210$ kJ
5.24. $\Delta H° = -141$ kJ
5.26. $\Delta H° = 85$ kJ
5.27. $\Delta H° = -285$ kJ
5.29. $\Delta H° = -141$ kJ
5.31. $\Delta H° = -254.47$
5.32. $\Delta H° = -155.1$ kJ
5.34. $\Delta H° = \frac{1}{2}\Delta H°_f\ (C) - \Delta H°_f\ (A) - \frac{1}{2}\Delta H°_f\ (B)$
5.35. $\Delta H° = 2\Delta H°_f\ (C) + \frac{1}{2}\Delta H°_f\ (D) - 3\Delta H°_f\ (A) - \Delta H°_f\ (B)$
5.37. $\Delta H° = -1010.2$ kJ
5.38. $\Delta H° = -361.9$ kJ
5.39. $\Delta H° = -198$ kJ
5.40. $\Delta H° = -74$ kJ

CHAPTER 6
6.2.

$\longrightarrow$
Na—O

6.3.

$\longrightarrow$
B—C

6.4.

$\longrightarrow$
S—O

6.5.

$\longrightarrow$
N—O

6.6.

$\longrightarrow$
B—I

6.7.

$\longrightarrow$
Mg—O

6.8.

$\longrightarrow$
Si—C

6.9.

$\longleftarrow$
O—H

6.10.

Cl — F

6.12. ionic

6.13. Polar covalent (although the electro-negativity difference between C and H is so small, that we often consider this to be a covalent bond).

6.14. covalent

6.15. ionic

6.16. polar covalent

6.17. ionic

6.18. ionic

6.19. polar covalent

6.20. polar covalent (although the electro-negativity difference between B and H is so small, that we often consider this to be a covalent bond).

6.22. +1

6.23. 0

6.24. −1

6.25. +1

6.26. −1

6.27. 0

6.29. −3

6.30. +3

6.31. −2

6.32. 0

6.33. −4

6.34. +2

6.35. Left is +1, Right is −3

6.36. +2

6.37. −2

6.39. +6

6.40. +6

6.41. +2

6.42. +3

6.43. +5

6.44. +4

6.46. +5

6.47. +6

6.48. +3

6.49. +3

6.50. +5

6.52. nitrogen is reduced; sulfur is oxidized

6.53. carbon is reduced; hydrogen is oxidized

6.54. chromium is reduced; chlorine is oxidized

6.55. nitrogen is reduced; zinc is oxidized

6.56. bromine is reduced; carbon is oxidized

CHAPTER 7

7.1. Consult the chart that precedes the problem.

7.3. :Br—Br:

7.4. :I—I:

7.5. H—F:

7.6. :Br—Cl:

7.7. H—I:

7.9. ⊖:C≡N:

7.10. :Cl—O:⊖

7.11. H—S:⊖

7.12. :N≡O:⊕

7.14. S

7.15. Al

7.16. C

7.17. C

7.18. S

7.19. Cl

7.20. Al

7.21. Se

7.22. I

7.23. As

7.24. Be

7.25. P

7.27.

$$O-Cl-O-H$$ with O above Cl

7.28.

$$H-O-Se-O-H$$ with O above Se

7.29.

$$O-Br-O-H$$ with O above Br and O below Br

7.30.
$$\begin{array}{c} O \\ \parallel \\ H-O-C-O-H \end{array}$$

7.31.
$$\begin{array}{c} O \\ \parallel \\ O-N-O-H \end{array}$$

7.32.
$$\begin{array}{c} H \\ | \\ O \\ | \\ H-O-P-O-H \\ | \\ O \end{array}$$

7.33.
$$\begin{array}{c} O \\ \parallel \\ O-C-O \end{array}$$

7.34.
$$\begin{array}{c} O \\ \parallel \\ O-N-O \end{array}$$

7.35.
$$\begin{array}{c} O \\ \parallel \\ H-O-P-O-H \\ | \\ O \end{array}$$

7.36.
$$\begin{array}{c} O \\ \parallel \\ O-C-O-H \end{array}$$

7.37.
$$\begin{array}{c} O \\ \parallel \\ O-I-O \end{array}$$

7.38.
$$\begin{array}{c} O \\ \parallel \\ O-S-O-H \\ \parallel \\ O \end{array}$$

7.40.
$$\begin{array}{c} :\!\ddot{O}\!: \\ \parallel \\ H-\ddot{O}-C-\ddot{O}-H \end{array}$$

7.41.
$$\ddot{O}=N-\ddot{O}-H$$

7.42.
$$\begin{array}{c} :\!\ddot{O}\!: \\ \parallel \\ :\!\ddot{C}l-C-\ddot{C}l\!: \end{array}$$

7.43.
$$\begin{array}{c} H \\ | \\ H-\overset{\oplus}{N}-H \\ | \\ H \end{array}$$

7.44.
$$\begin{array}{c} H \\ | \\ H-\overset{\ominus}{Al}-H \\ | \\ H \end{array}$$

7.45.
$$\begin{array}{c} :\!\ddot{C}l\!: \\ | \\ :\!\ddot{C}l-\overset{\oplus}{P}-\ddot{C}l\!: \\ | \\ :\!\ddot{C}l\!: \end{array}$$

7.46.
$$\ddot{S}=C=\ddot{S}$$

7.47.
$$\begin{array}{c} :\!\ddot{F}\!: \\ | \\ :\!F-Si-F\!: \\ | \\ :\!\ddot{F}\!: \end{array}$$

7.48.
$$\begin{array}{c} :\!\ddot{F}\!: \\ | \\ :\!F-\overset{\ominus}{B}-F\!: \\ | \\ :\!\ddot{F}\!: \end{array}$$

7.50.
$$\begin{array}{c} :\!\ddot{C}l\!: \\ | \\ :\!\ddot{C}l-B-\ddot{C}l\!: \end{array}$$

7.51.
$$\begin{array}{c} :\!\ddot{C}l\!: \\ | \\ :\!\ddot{C}l-\overset{\ominus}{Al}-\ddot{C}l\!: \\ | \\ :\!\ddot{C}l\!: \end{array}$$

7.52.
$$\begin{array}{c} H \\ | \\ H-B-H \end{array}$$

7.54.
$$:\!\overset{\ominus}{\ddot{O}}-\overset{\oplus}{\ddot{C}l}-\ddot{O}-H \quad \underline{\textit{or}} \quad \ddot{O}=\ddot{C}l-\ddot{O}-H$$

7.55.
$$\begin{array}{c} :\!\overset{\ominus}{\ddot{O}} \\ | \\ :\!\ddot{O}-\underset{\oplus 2}{I}-\ddot{O}-H \end{array} \quad \underline{\textit{or}} \quad \begin{array}{c} :\!\ddot{O}\!: \\ \parallel \\ \ddot{O}=I-\ddot{O}-H \end{array}$$

7.56.
$$:\!\overset{\ominus}{\ddot{O}}-\underset{\oplus 2}{S}-\overset{\ominus}{\ddot{O}}\!: \quad \underline{\textit{or}} \quad \ddot{O}=S=\ddot{O}$$

7.57.
$$\begin{array}{c} :\!\overset{\ominus}{\ddot{O}}\!: \\ | \\ :\!\overset{\ominus}{\ddot{O}}-\underset{\oplus 3}{S}-\overset{\ominus}{O} \end{array} \quad \underline{\textit{or}} \quad \begin{array}{c} :\!\ddot{O}\!: \\ \parallel \\ \ddot{O}=S-\overset{\ominus}{O} \end{array}$$

7.58.
$$\begin{array}{c} :\!\overset{\ominus}{\ddot{O}}\!: \\ | \\ :\!\overset{\ominus}{\ddot{O}}-\underset{\oplus 3}{Br}-\ddot{O}-H \\ | \\ :\!\overset{\ominus}{\ddot{O}}\!: \end{array} \quad \underline{\textit{or}} \quad \begin{array}{c} :\!\ddot{O}\!: \\ \parallel \\ \ddot{O}=Br-\ddot{O}-H \\ \parallel \\ :\!\ddot{O}\!: \end{array}$$

7.59.

7.61.

7.62.

CHAPTER 8

8.2. linear
8.3. trigonal planar
8.4. tetrahedral
8.5. octahedral
8.6. trigonal bipyramidal
8.7. trigonal planar
8.9. tetrahedral
8.10. trigonal bipyramidal
8.11. tetrahedral
8.12. octahedral
8.13. tetrahedral
8.16. trigonal pyramidal
8.17. trigonal planar
8.18. linear
8.19. tetrahedral
8.20. tetrahedral
8.21. T-shape
8.22. seesaw
8.23. square planar
8.24. trigonal pyramidal
8.25. linear
8.26. trigonal pyramidal
8.27. square pyramidal
8.28. seesaw
8.29. trigonal bipyramidal
8.30. octahedral
8.31. linear
8.33. trigonal planar

8.34. linear
8.35. linear
8.37. Trigonal bipyramidal. Equatorial groups are all 120° apart from each other, and axial groups are 90° apart from all equatorial groups.
8.38. Trigonal pyramidal. All angles are slightly less than 109.5°.
8.39. Seesaw (based on trigonal bipyramidal arrangement). One equatorial site is taken by the lone pair. The other two equatorial groups should be slightly less than 120° apart from each other (because of the effect of the lone pair), and the axial groups should be 90° apart from the equatorial groups.
8.40. Trigonal bipyramidal. Equatorial groups are all 120° apart from each other, and axial groups are 90° apart from all equatorial groups.
8.42. sp^3
8.43. sp^3d^2
8.44. sp^3 (this is just a prediction; experimental evidence suggests otherwise)
8.45. sp^3
8.46. sp^3d
8.47. sp^2
8.48. sp^3
8.49. sp^3d
8.50. sp^3d^2
8.51. sp^3d
8.52. sp^3d^2
8.53. sp^3d
8.54. sp^3d^2
8.55. sp^2
8.56. sp
8.57. sp^3
8.58. sp^3d
8.59. sp^3d
8.60. sp
8.61. sp^3d
8.62. sp^3d^2
8.63. sp^3
8.64. sp
8.65. sp^2
8.66. sp^2
8.67. sp^3d^2
8.68. sp^3
8.69. sp^3

CHAPTER 9

9.2. 10^{26}

9.3. 10^{20}

9.4. 10^{69}

9.5. 10^{13}

9.6. 10^{14}

9.7. 10^{9}

9.8. 10^{12}

9.9. 10^{8}

9.10. 10^{-26}

9.11. 10^{22}

9.14. $x = \log 572.4 = 2.7577$

9.15. $x = \log 1 = 0$

9.16. $x = \log 52 = 1.72$

9.17. $x = 10,000,000 = 7$

9.19. $y = e^{2.78}$

9.20. $K = e^{-\Delta G^\circ / RT}$

9.22. $x = \ln 275 = 5.617$

9.23. $x = \ln 1 = 0$

9.25. 2.5400

9.26. 5.8485

9.27. 1.39804

9.28. -5.65

9.29. -4.2

9.30. 3.039955×10^{-5}

9.32. $\log (V_1) - \log (V_2)$

9.33. $\ln (P_1) + \ln (P_2)$

9.34. $d \times [\ \log (a) + \log (b) + \log (c)] = [d \times \log (a)] + [d \times \log (b)] + [d \times \log (c)]$

9.35. $d \times [\ \ln (a) + \ln (b) - \ln (c)] = [d \times \ln (a)] + [d \times \ln (b)] - [d \times \ln (c)]$

9.37. $\ln (P_2/P_1)$

9.38. $\ln (P_2 \times P_1)$

9.39. $\log (K_1/K_2)$

9.40. $\log \left(\dfrac{a \times c}{b} \right)$

APPENDIX

A.2. $Al(OH)_3 + 3\ HClO_4 \longrightarrow Al(ClO_4)_3 + 3\ H_2O$

A.3. $Cu(NO_3)_2\ (s) + 2\ KOH\ (aq) \longrightarrow Cu(OH)_2\ (s) + 2\ KNO_3\ (aq)$

A.4. $2\ C_2H_2Cl_4 + Ca(OH)_2 \longrightarrow 2\ C_2HCl_3 + CaCl_2 + 2\ H_2O$

A.5. $NaCl\ (aq) + 3\ H_2O\ (l) \longrightarrow NaClO_3\ (aq) + 3\ H_2\ (g)$

A.6. $3\ Al\ (s) + 3\ NH_4ClO_4\ (s) \longrightarrow Al_2O_3\ (s) + AlCl_3\ (s) + 3\ NO\ (g) + 6\ H_2O\ (l)$

A.7. $C_3H_8\ (g) + 5\ O_2\ (g) \longrightarrow 3\ CO_2\ (g) + 4\ H_2O\ (g)$

A.8. $CaSiO_3\ (s) + 6\ HF\ (g) \longrightarrow SiF_4\ (g) + CaF_2\ (s) + 3\ H_2O\ (l)$

A.9. $2\ C_3H_6\ (g) + 2\ NH_3\ (g) + 3\ O_2\ (g) \longrightarrow 2\ C_3H_3N\ (g) + 6\ H_2O\ (l)$

INDEX